Lecture Notes in Mathematics

Edited by A. Dold and B. Eckmann

453

Logic Colloquium

Symposium on Logic Held at Boston, 1972–73

Edited by R. Parikh

Springer-Verlag
Berlin · Heidelberg · New York 1975

Prof. Dr. Rohit Parikh
Department of Mathematics
Boston University
College of Liberal Arts
Boston
Massachusetts 02215/USA

Library of Congress Cataloging in Publication Data
Main entry under title:

Logic Colloquium.

(Lecture notes in mathematics ; 453)
Based on talks at the Boston Logic Colloquium in 1972-73.
Includes bibliographies and index.
1. Logic, Symbolic and mathematical--Congresses.
I. Parikh, Rohit, 1936- II. Logic Colloquium, Boston, 1972-1973. III. Series: Lecture notes in mathematics (Berlin) ; 453.
QA3.L28 no. 453 [QA9] 510'.8s [511'.3] 75-11528

AMS Subject Classifications (1970): 02B99, 02C10, 02C99, 02D99, 02F27, 02G05

ISBN 3-540-07155-5 Springer-Verlag Berlin · Heidelberg · New York
ISBN 0-387-07155-5 Springer-Verlag New York · Heidelberg · Berlin

Offsetdruck: Julius Beltz, Hemsbach/Bergstr.

PREFACE

The papers in this volume originated as talks given at the Boston Logic Colloquium during the year 1972-73. However, some of them contain more recent developments not originally included in the talks. They are all technical papers in mathematical logic but with a strong foundational interest. Thus they should also be of interest to philosophers.

Eleven talks were actually given at the colloquium. However, the talks by Abraham Robinson and Alexander Yessenin-Volpin were expository and the one by Ivor Grattan-Guinness was historical. The eleventh talk, by John Myhill, on constructive set theory, will appear in the Journal of Symbolic Logic.

Funds for the colloquium were provided by the graduate school of Boston University through the Boston Colloquium for the Philosophy of Science.

The colloquium was a joint effort by many and I am grateful to the following for their help. George Berry, Robert Cohen, David Ellerman, James Geiser, Ilona Webb, Judson Webb, and Marx Wartofsky.

August 12, 1974 Rohit Parikh

CONTENTS

COMBINATORIAL FUNCTORS

J.N. Crossley and Anil Nerode

Algebra and model theory deal with properties preserved under isomorphism. We first of all develop the theory of a new sort of continuous functor and then effectivize everything. The techniques developed greatly facilitate the study of properties preserved under effective isomorphisms.

The work outlined here will appear in a much more extensive treatment in our forthcoming monograph [Crossley & Nerode 1974].* It represents the latest developments in the theory of combinatorial functors and recursive equivalence types which has been extensively developed principally by Dekker, Myhill, Ellentuck, Nerode but also by many others (see the bibliography of Crossley [1970]). We are grateful to all these for giving us a foundation on which to build. We also acknowledge financial support from Cornell University, U.C.L.A., Monash University and The National Science Foundation from 1966 on. We are also grateful to Liz Wachs and Bill Gross for many improvements to the original version.

Introductions to the special cases of recursive equivalence types of sets are most readily provided by Dekker's useful little book [1966] and Dekker and Myhill's pioneering monograph [1960]. For the set-theoretic approach see Ellentuck [1965]. A general survey to 1969 is given in Crossley [1970].

Peter Aczel's regrettably unpublished dissertation [1966] was the first hint in print that a functorial point of view is profitable in RET's.

---

* The book has already appeared

1. Categories

For simplicity we consider a fixed but arbitrary category $\mathbb{C}$ which arises in the following way. There is a (countable) structure $\mathfrak{U} = \langle U, f_i \rangle_{i \in I}$ where each f_i is a (partial) function from $X^n U$ to U for some n depending on i and the universe U is (a subset of) the natural numbers E. (The extension from algebras to structures is clear and will be assumed later but for the present we prefer a purely algebraic approach.)

The objects of $\mathbb{C}$, denoted by $\mathfrak{U}, \mathfrak{B}, \ldots,$ are subsystems of U and the morphisms of $\mathbb{C}$ are one-one maps between objects of $\mathbb{C}$ preserving the structure of $\mathfrak{U}$. By a map we mean a triple $p = (D,f,C)$ where $D,C \subseteq E$ and p is a single valued set of ordered pairs, $D = \text{dom } p$ is the set of first elements of ordered pairs in p, $C = \text{codom } p$ contains the set of second elements of ordered pairs in p and we write $f = \text{graph } p$. If $p_o = (D_o,f_o,C_o)$ and $p_1 = (D_1,f_1,C_1)$ are maps then we define $p_o \cap p_1 = (D_o \cap D_1, f_o \cap f_1, C_o \cap C_1)$ if, but only if, this triple is a map in our sense. Similarly $p_o \subseteq p_1$ iff $p_o \cap p_1 = p_o$.

We require $\mathbb{C}$ to satisfy the following conditions. (The numbering is taken from Crossley & Nerode [1974].)

(1.3) Inverses. If p is a morphism of $\mathbb{C}$ and p, as a map, has an inverse p^{-1}, then p^{-1} is a morphism of $\mathbb{C}$.

(1.4) Inclusions. We write $\text{Ob}(\mathbb{C})$ for the objects of $\mathbb{C}$. If $\mathfrak{A}, \mathfrak{B} \in \text{Ob}(\mathbb{C})$ and $\mathfrak{A}$ is a subset of $\mathfrak{B}$ then the inclusion map $\mathfrak{A} \to \mathfrak{B}$ is a morphism of $\mathbb{C}$.

Since $\mathfrak{U}$ is an algebra the intersection of subalgebras is again a subalgebra and similarly for monomorphisms. We require

(1.5) Intersection. If a set of morphisms of $\mathbb{C}$ has an intersection (as a set of maps) then that intersection is in $\mathbb{C}$.

(1.6) Directed unions. The union of any set of morphisms of $\mathbb{C}$, directed under inclusion is a morphism of $\mathbb{C}$.

This is a variation on the familiar fact that the directed union of algebras (of a given sort) is an algebra. As usual in category theory we identify objects with the identity morphisms on them.

Any object $\mathfrak{A}$ in $\mathbb{C}$ is the universe of a subalgebra. We ambiguously write $\mathfrak{A}$ for this subalgebra.

(1.7) Subcategory. The finitely generated objects in $\mathbb{C}$ form a full subcategory ${}^{o}\mathbb{C}$.

(1.8) Restriction. If p is a morphism of $\mathbb{C}$, $\mathfrak{A} \in \mathrm{Ob}(\mathbb{C})$ and $\mathfrak{A} \subseteq \mathrm{dom}\ p$ then the bijection $q \subseteq p$ with domain $\mathfrak{A}$ is a morphism of $\mathbb{C}$.

We write $p|\mathfrak{A}$ for this morphism.

Note that if $\mathfrak{A} \in {}^{o}\mathbb{C}$ and $p \in \mathbb{C}$ then $p|\mathfrak{A} \in {}^{o}\mathbb{C}$.

One of our main techniques will involve extending functors from finitely generated subalgebras to infinitely generated ones. So we require our objects and morphisms to satisfy the next condition.

(1.9) Approximation. If p is a morphism of $\mathbb{C}$ then the morphisms p' of ${}^{o}\mathbb{C}$ with $p' \subseteq p$ form a directed set (under inclusion) with union p.

(1.9) allows us to approximate "large" objects or morphisms. The next condition, when used in conjunction with (1.6) allows us to build up large objects (and morphisms). We say that a set of morphisms $p, q, \ldots$ is compatible in $\mathbb{C}$ if there is a morphism r in $\mathbb{C}$ such that $p \subseteq r$, $q \subseteq r, \ldots$.

(1.10) Compatibility. If p, q are morphisms of ${}^{\circ}\mathbb{C}$ then there is (under $\subseteq$) a least morphism r in $\mathbb{C}$ such that $p, q \subseteq r$ and, moreover, $r \in {}^{\circ}\mathbb{C}$.

Induction immediately yields a generalization of (1.10) to any finite set of morphisms.

By (1.6) and (1.10) if $\{p_i : i \in I\}$ is any compatible set of morphisms then there is a least morphism $q \supseteq p_i$ for all $i \in I$. We write $q = \bigvee\{p_i : i \in I\}$ and similarly for objects.

Finally we note that the morphisms in ${}^{\circ}\mathbb{C}$ are finitely generated in the sense that for each $r \in {}^{\circ}\mathbb{C}$ if $r = (\mathfrak{A}, f, \mathfrak{B})$ and A, B are sets of generators of the subalgebras $\mathfrak{A}$, $\mathfrak{B}$ then any morphism containing the map $(A, f|A, f(A) \cup B)$ is an extension of r.

A category of subalgebras of the above type is said to be an appropriate category.

It is useful to mention here that if $\mathbb{C}_0, \ldots, \mathbb{C}_n$ are appropriate then so is $\mathbb{C}_0 \times \ldots \times \mathbb{C}_n$ whose objects (morphisms) are $(n+1)$-tuples $(p_0, \ldots, p_n)$ where $p_i \in \mathbb{C}_i$ and all operations are performed co-ordinatewise. We shall not draw attention to this fact in the theory below but shall use it implicitly in the examples.

Example 1.1: The category $\$$ (sets). The objects are sets of natural numbers and the morphisms one-maps between sets of natural numbers. ${}^{\circ}\$$ is the full subcategory whose objects are finite set of natural numbers.

Example 1.2. The category $\mathbb{L}$ (linear orderings). Objects are sets of rationals, morphisms are one-one order preserving maps, ${}^{\circ}\mathbb{L}$ is the full subcategory of finite sets.

Example 1.3. The category $\mathbb{V}$ (vector spaces). The objects are those subsets of a fixed countably infinite dimensional vector space over a fixed field K which are subspaces. The morphisms are vector space linear transformations between objects. ${}^{\circ}\mathbb{V}$ is the full subcategory of finite dimensional subspaces.

Example 1.4. The category $\mathbb{B}$ (Boolean algebras). The objects are those subsets of a fixed countable atomless Boolean algebra which are non-empty and closed under the Boolean operations (subalgebras). The morphisms are Boolean monomorphisms between objects and ${}^{\circ}\mathbb{B}$ is the full subcategory of finite Boolean subalgebras.

2. Functors.

From now on we shall assume that all categories considered are appropriate in the sense of section 1.

We use the word 'functor' in a more concrete way than is often the case. A functor $F: \mathbb{C} \to \mathbb{C}'$ is a function which assigns to each morphism $p \in \mathbb{C}$ a morphism $Fp \in \mathbb{C}'$ such that $F(pq) = F(p)F(q)$ and, moreover, if $p: \mathfrak{A} \to \mathfrak{B}$ is an inclusion morphism, then $F(p) : F(\mathfrak{A}) \to F(\mathfrak{B})$ is the inclusion morphism. (Note that this condition means $F(p)$ is an identity if p is.)

We shall be concerned with combinatorial functors which are a subclass of the continuous ones in the following topology. We topologize $\mathbb{C}$. Set

$$U(p) = \{q \in \mathbb{C} : p \subseteq q\}.$$

The weak topology on $\mathbb{C}$ is the smallest topology with $U(p)$ open for all $p \in {}^{\circ}\mathbb{C}$. By the compatibility condition (1.10) the $U(p)$ for $p \in {}^{\circ}\mathbb{C}$ together with the empty set form a base for the weak topology. We call this base the standard base.

Definition 2.1. A functor $F : \mathbb{C} \to \mathbb{C}'$ is said to be combinatorial if (i) F is continuous on morphisms, that is $F(p) = \cup F(q) : q \subseteq p \;\epsilon\; q \in {}^{\circ}\mathbb{C}\}$ and (ii) for any $\mathfrak{B}' \in {}^{\circ}\mathbb{C}'$, if $\mathfrak{B}' \subseteq F(\mathfrak{V})$ for some $\mathfrak{V}$ then there exists $\mathfrak{A} \in {}^{\circ}\mathbb{C}$ such that for all $\mathfrak{L} \in \mathbb{C}$

(*) $$\mathfrak{B}' \subseteq F(\mathfrak{L}) \quad \text{iff} \quad \mathfrak{A} \subseteq \mathfrak{L} \quad .$$

In this case we write $\mathfrak{A} = F^{\leftarrow}(\mathfrak{B}')$.

Theorem 2.2. A functor $F : \mathbb{C} \to \mathbb{C}'$ continuous on morphisms is combinatorial iff the inverse image of a standard base set in Ob C' is a standard base set in Ob C.

Corollary 2.3. Combinatorial functors are continuous, preserve inclusions and intersections and are closed under composition.

Note that to establish closure under composition in the many variables case the compatibility condition (1.10) is required.

The next theorem is most useful.

Theorem 2.4. Let $F : {}^{\circ}\mathbb{C} \to \mathbb{C}'$ be a combinatorial functor then there is unique extension $G : \mathbb{C} \to \mathbb{C}'$ of F which is a combinatorial functor.

The extension G is given by

$$G(\mathfrak{B}) = \cup\{F(\mathfrak{A}) : \mathfrak{A} \subseteq \mathfrak{B} \;\epsilon\; \mathfrak{A} \in {}^{\circ}\mathbb{C}\}$$

and similarly for morphisms.

The following characterization is also useful.

Theorem 2.5. Suppose ${}^{\circ}\mathbb{C}$ is closed under arbitrary intersections, Let $F : \mathbb{C} \to \mathbb{C}'$ be a functor continuous on morphisms then F is combinatorial iff F preserves arbitrary intersections of objects.

Theorem 2.6. Suppose ${}^{\circ}\mathbb{C}$ is closed under arbitrary intersections. Then $F : \mathbb{C} \to \mathbb{C}'$ is combinatorial iff F is continuous on morphisms

$$v_p(f) = \lim_{n\to\infty} S_p^{s_n}(f)$$

$$= \lim_{\alpha\downarrow 0} \lim_{n\to\infty} S_{p,\alpha}^{s_n}(f) + \sum_{k=1}^{\infty} |f(y_k) - f(x_k)|^p \quad ;$$

le premier terme, noté $v_p^*(f,F)$, est appelé p-variation fine de f sur F ; le second est appelé p-variation grossière de f sur F. Nous allons donner ici une expression explicite de la p-variation fine : $v_p^*(f,F)$.

DEFINITION 1. Pour toute fonction $f : T \to \mathbb{R}$ et toute partie A de $\overline{\mathbb{R}}$, on appelle p-variation fine de f sur A le nombre

(12) $$v_p^*(f,A) = \lim_{\alpha\downarrow 0} \sup_{\substack{s\subset A\cap T \\ s \text{ fini}}} \sum_{i=1}^{n-1} (|f(t_{i+1}) - f(t_i)|^p \wedge \alpha) ,$$

où $s = (t_1, t_2, \ldots, t_n)$. En particulier on appelle p-variation fine de f le nombre

(13) $$v_p^*(f) = v_p^*(f,T).$$

Le théorème suivant donne deux autres expressions de la p-variation fine.

THEOREME 6. Soit une fonction $f \in \mathcal{B}_p(T)$, dont $f' \circ g$ est la factorisation canonique. Pour toute partie A de T, $[x_k , y_k]$ $(k\in\mathbb{N})$ désignant les intervalles contigus à $\overline{g(A)}$,

(14) $$v_p^*(f,A) = \inf_{s\subset A} \sum_{i=0}^{n} v_p(f ,]t_i , t_{i+1}[\cap A)$$

(15) $$v_p^*(f,A) + \sum_{k=1}^{\infty} |f'(y_k) - f'(x_k)|^p = \inf_{s\subset A} \sum_{i=0}^{n} v_p(f , [t_i,t_{i+1}]\cap A) ,$$

avec $s = (t_1, t_2, \ldots, t_n)$, $t_o = \inf A$, $t_{n+1} = \sup A$.

If $x \in \text{dom}\oplus p$ then $x \in f_0(E)$ or $x \in f_1(E)$ but not both. If $x \in f_0(E)$ then $x = f_0(y_0)$ for some natural number y_0. Set $\oplus p(x) = \oplus p(f_0(y_0)) = f_0(p_0(y_0))$ and similarly if $x \in f_1(E)$ and $x = f_1(y)$ set

$$\oplus p(x) = \oplus p(f_1(y_1)) = f_1(p_1(y_1)).$$

Clearly, when p is an inclusion morphism $A \to B$, $\oplus p$ is an inclusion morphism $\oplus A \to \oplus B$.

We verify that $\oplus$ is a functor. Suppose p,q are morphisms of $\$ \times \$$ such that codom p = dom q. Then for i = 0.1 codom p_i = domq_i and

$$\oplus(q o p)f_i(x) = f_i(y) \text{ iff } (q o p)_i(x) = y$$

iff $(q_i \circ p_i)(x) = y$ by definition of $\circ$ in $\$ \times \$$

$$\begin{aligned} \text{iff } \oplus q(f_i p_i(x)) &= f_i q_i p_i(x) \\ &= f_i(y) \end{aligned}$$

$$\text{iff } \oplus q(\oplus p(f_i(x)) = f_i(y).$$

Finally we check that $\oplus$ is combinatorial. If A is an object of ${}^{o}\$$ set

$$(\oplus^{\leftarrow} A)_i = \{x : f_i(x) \in A\}.$$

(Recall that $(B)_i$ is the i-th co-ordinate of B and $\oplus^{\leftarrow} A \in \$ \times \$$.) Then $x \in \oplus A$ implies $(\oplus^{\leftarrow}\{x\})_i \subseteq A_i$ $(i = 0,1)$ and if $(\oplus^{\leftarrow}\{x\})_i \subseteq A_i$ then from the definition of $\oplus$, $x \in \oplus(A_0, A_1)$. Similarly with $\{x\}$ replaced by a set B. Since $\oplus$ is clearly continuous on morphisms $\oplus$ is now combinatorial.

<u>Example 2.2.</u> <u>Cardinal multiplication</u> (cf. Sierpinski [1958] p.135). We define a combinatorial functor $\otimes : \$ \times \$ \to \$$. This functor underlies multiplication of countable cardinals.

Let $j : E \times E \to E$ be a bijection (which may be assumed to be recursive).

For objects A of $S \times S$ set

$$\otimes A = \{j(x,y) : x \in A_0 \in y \in A_1\}.$$

This defines the functor on objects so now we must define it on morphisms p of $\$ \times \$$. Set $\mathrm{dom}\otimes p = \mathrm{dom}\, p_0 \otimes \mathrm{dom} p_1$, $\mathrm{codom} \otimes p = \mathrm{codom}\, p_0 \otimes \mathrm{codom}\, p_1$. $\otimes p\ j(x,y)$ is to be defined when $p_0(x)$, $p_1(y)$ are defined and then

$$\otimes p(j(x,y)) = j(p_0(x),p_1(y)).$$

We leave the reader to check that $\otimes$ is combinatorial noting only that

$$(\otimes^{\leftarrow} A)_0 = \{x : \exists\, y(j(x,y) \in A)\}$$

and

$$(\otimes^{\leftarrow} A)_1 = \{y : \exists\, x(j(x,y) \in A)\}.$$

<u>Example 2.3</u>. <u>Ordinal addition</u>. We define a combinatorial functor $\oplus: \mathbb{L} \times \mathbb{L} \to \mathbb{L}$ underlying the addition of countable order types. Let $\mathfrak{A} = (Q,<)$ be the rational numbers. Let $f^0, f^1 : Q \to Q$ be order preserving injections such that $f^0(Q)$ and $f^1(Q)$ are disjoint and exhaust Q and every element of $f^0(Q)$ precedes every element of $f^1(Q)$.

$f^0(Q)$ | $f^1(Q)$

These injections exist because the ordering pictured above is a countable dense unbordered linear ordering and therefore, by a theorem of Cantor, (order) isomorphic to Q. Further, because of the constructive character of the proof f^0, f^1 may be chosen to be general recursive.

We now identify Q with E and endow E with the induced dense ordering $\prec$. Then if we use exactly the same functor as for cardinal addition we obtain the required combinatorial functor.

Example 2.4. Ordinal multiplication. We define a combinatorial functor $\otimes : \mathbb{L} \times \mathbb{L} \to \mathbb{L}$ which underlies ordinal multiplication. Let $j : Q \times Q \to Q$ be a bijection such that for all $(a_o, a_1), (b_o, b_1)$ in $Q \times Q$, $j(a_o, a_1) < j(b_o, b_1)$ if, and only if, $a_1 < b_1$ or $(a_1 = b_1$ & $a_o < b_o)$. Again j exists since $Q \times Q$ with the lexicographic ordering is a countable dense unbordered linear ordering and therefore isomorphic to Q. Again also j may be chosen to be recursive. Now use the same functor $\otimes$ as for cardinal multiplication and show that $\otimes$ is indeed a combinatorial functor.

In the same way we can define (weak) cardinal exponentiation and ordinal exponentiation following Sierpinski [1958].

Example 2.5. Let $V : \mathbb{S} \to \mathbb{V}$ be defined as follows. Identify $\mathbb{S}$ with the category of one-one maps between subsets of a countably infinite set A. Let f be a one-one map from A onto a basis for a vector space U over a field K and let $\mathbb{V}$ be the category of vector subspaces of U with one-one linear transformations between them. Then V assigns to each object $\mathfrak{A}$ of $\mathbb{S}$ the object $V(\mathfrak{A})$ generated by $\{f(a) : a \in \mathfrak{A}\}$ and to each morphism p of $\mathbb{S}$ the morphism such that the following diagram commutes

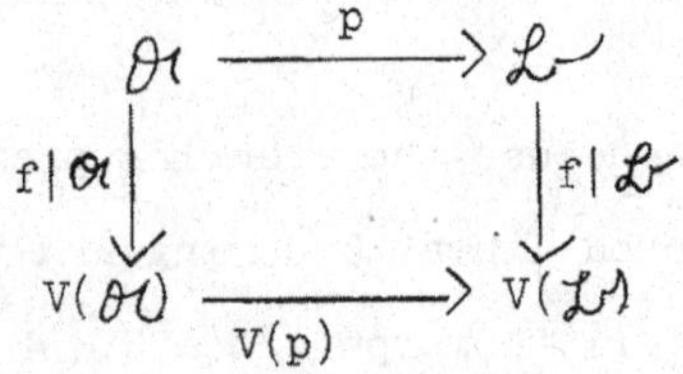

$$\begin{array}{ccc} \mathfrak{A} & \xrightarrow{p} & \mathfrak{B} \\ {\scriptstyle f|\mathfrak{A}}\downarrow & & \downarrow{\scriptstyle f|\mathfrak{B}} \\ V(\mathfrak{A}) & \xrightarrow[V(p)]{} & V(\mathfrak{B}) \end{array}$$

V is a combinatorial functor.

3. Model Theory. Now we turn to model theory for it is from certain recursive models that we get out best examples for the theory of recursive combinatorial functors and the model theory also enables us in some cases to relate combinatorial functors to number theoretic functions.

We consider a fixed $\aleph_0$-categorical theory T which has an infinite model $\mathfrak{M}$. We may assume that every formula is equivalent to an atomic formula (or else add new predicate letters and axioms to make this so).

For a formula ϕ with $v_0, v_1, \ldots, v_n$ free then if

$$\mathfrak{M} \vdash \exists v_n \phi(v_0, \ldots, v_n) \quad [a_0, \ldots, a_{n-1}]$$

either, for some positive natural number k,

$$\mathfrak{M} \vdash \exists !^k v_n \phi(v_0, \ldots, v_n) \; [a_0, \ldots, a_{n-1}]$$

(where $\exists !^k$ means "there are exactly k")
or there are infinitely many elements b in $|\mathfrak{M}|$ such that $\mathfrak{M} \vdash \phi[a_0, \ldots, a_{n-1}, b]$. If there is some formula ϕ such that the former case holds we say that an element b such that $\mathfrak{M} \vdash \phi[a_0, \ldots, a_{n-1}, b]$ is algebraic over $\{a_0, \ldots, a_{n-1}\}$. b is algebraic over a set $A \subseteq |\mathfrak{M}|$ if b is algebraic over some finite subset of A. The algebraic closure $cl(A)$ of a set $A \subseteq |\mathfrak{M}|$ is the smallest subsystem $\mathfrak{L} \subseteq \mathfrak{M}$ such that $A \subseteq |\mathfrak{L}|$ and if b is algebraic over $\mathfrak{L}$, then $b \in |\mathfrak{L}|$.

A model $\mathfrak{M}$ is said to be recursively presented if the universe of $\mathfrak{M}$ is an initial segment of the natural numbers and the satisfaction relation "a satisfies ϕ in $\mathfrak{M}$" is recursive (where a is a finite sequence of elements of $|\mathfrak{M}|$ and ϕ a formula). If T is a decidable theory then T has a recursively presented model as can easily be seen by effectivising a standard Henkin-style completeness proof.

We are now almost ready to define categories which are suitable for developing the theory of recursive equivalence and recursive combinatorial functors.

A category $\mathbb{C} \equiv \mathbb{C}(\mathfrak{M})$ is said to be <u>suitable</u> if $\mathbb{C}$ is the category of elementary monomorphisms between algebraically closed subsets of a recursively presented infinite model $\mathfrak{M}$ of an $\aleph_0$-categorical theory T which is decidable, has decidable atoms and is such that algebriac closure is fully effective. ${}^{\circ}\mathbb{C}(\mathfrak{M})$ is the full subcategory of finite objects and morphisms. Note that any suitable category satisfies all the conditions in section 1. To say T has decidable atoms means that given a formula ϕ with at most $v_0, \ldots, v_{n-1}$ free then we can effectively decide whether ϕ is an atom of $B_n(T)$ or not. To say algebraic closure is fully effective means that given an explicit index of a finite set $A \subseteq |\mathfrak{M}|$ we can uniformly effectively compute an explicit index of the algebraic closure, $cl(A)$, of A. Note that T being $\aleph_0$-categorical means $cl(A)$ is finite if A is finite.

The examples of Section 1 are (up to isomorphism) suitable categories (if the field K in 1.3 is finite). For 1.1 T is the theory of an infinite set with equality, for 1.2 that of a dense unbordered linear ordering, for 1.3 that of an infinite dimensional vector space over K and for 1.4 that of an atomless Boolean algebra. So we shall now use $\mathbb{S}$, $\mathbb{L}$, $\mathbb{V}$ and $\mathbb{B}$ for the categories which arise as $\mathbb{C}(\mathfrak{M})$ for $\mathfrak{M}$ the recursively presented model of the appropriate theory T.

In the case of $\mathbb{S}$ and $\mathbb{V}$ but not $\mathbb{L}$ or $\mathbb{B}$ we have the notions of dimension and basis akin to those for vector spaces. This is also the case in many other suitable categories. (In the case of other categories, in particular $\mathbb{B}$, there is a notion of dimension but this does not have sufficiently strong properties for our purposes.) We shall not however give the formal criteria here but just treat $\mathbb{S}$ and $\mathbb{V}$ when considering categories with dimension. In $\mathbb{S}$ dimension of A = cardinal of A and every element is a basis element. In $\mathbb{V}$ dimension is the usual vector space dimension and basis has the usual meaning.

4. Combinatorial Functions. For categories with dimension we define a subclass of combinatorial functors.

A map $G : Ob({}^{\circ}\mathbb{C}) \to \mathcal{P}(|\mathfrak{M}'|)$ is said to be a precombinatorial operator if

1) $\mathfrak{A} \neq \mathfrak{A}'$ implies $G(\mathfrak{A}) \cap G(\mathfrak{A}') = \emptyset$ for $\mathfrak{A}, \mathfrak{A}' \in {}^{\circ}\mathbb{C}(\mathfrak{M})$,

2) if some $p : \mathfrak{A} \to \mathfrak{A}'$ is an isomorphism in ${}^{\circ}\mathbb{C}(\mathfrak{M})$ then card $G(\mathfrak{A})$ = card $G(\mathfrak{A}')$ and

3) $\cup\{G(\mathfrak{A}) : \mathfrak{A} \in {}^{\circ}\mathbb{C}\}$ is a subset of a basis for $\mathfrak{M}'$.

A combinatorial functor $F : \mathbb{C}(\mathfrak{M}) \to \mathbb{C}(\mathfrak{M}')$ is said to be strict if there is a precombinatorial operator G inducing F in the sense that

$$F(\mathfrak{A}) = c\ell \cup \{G(\mathfrak{A}') : \mathfrak{A}' \subseteq \mathfrak{A} \in \mathfrak{A}' \in {}^{\circ}\mathbb{C}\}.$$

All $\mathbb{S}$-valued combinatorial operators are strict and

$$G(\mathfrak{A}) = F(\mathfrak{A}) - \cup\{F(\mathfrak{A}') : \mathfrak{A}' \in {}^{\circ}\mathbb{C}(\mathfrak{M}) \ \& \ \mathfrak{A}' \subsetneq \mathfrak{A}\}$$

is the precombinatorial operator inducing F.

Theorem 4.1. Strict combinatorial functors are closed under composition.

A combinatorial functor F is said to be finitary if $\mathfrak{A}$ finite implies $F(\mathfrak{A})$ finite. Each finitary strict combinatorial functor $F : \mathbb{C} \to \mathbb{C}'$ where $\mathbb{C}, \mathbb{C}'$ are suitable and $\mathbb{C}, \mathbb{C}'$ have dimension induces a number theoretic function $F^{\#}$ given by

$$F^{\#} \dim \mathfrak{A} = \dim F(\mathfrak{A}) \quad \text{for } \mathfrak{A} \text{ in } {}^{\circ}\mathbb{C}$$

where dim denotes dimension in the appropriate category. Such an induced number theoretic function is said to be a <u>strict</u> $\mathbb{C}$ to $\mathbb{C}'$ <u>combinatorial function</u>. If $\mathbb{C} = \mathbb{C}'$ we call the function strict $\mathbb{C}$-combinatorial.

<u>Corollary 4.2</u>. Strict combinatorial functions are closed under composition.

Let $[{n \atop i}]_{\mathbb{C}}$ denote the number of i-dimensional subobjects of an n-dimensional object in the category $\mathbb{C}$ (which is assumed to have dimension). We omit the subscript where there is no ambiguity.

<u>Theorem 4.3</u>. (Myhill normal form). A function $f : E \to E$ is strict combinatorial iff there is a function $c : E \to E$ such that for all n in E

$$f(n) = \Sigma\, c(i)[{n \atop i}].$$

The function c given by Theorem 4.3 is said to be the <u>Stirling coefficient function</u> for f. It is easy to show that the Stirling coefficient function is unique.

<u>Example 4.1</u>. In $\mathbb{S}$, $[{n \atop i}] = ({n \atop i}) = \frac{n!}{i!(n-i)!}$.

Here any $f : E \to E^* = \{0, \pm 1, \pm 2, \ldots\}$ has a unique expansion $f(n) = \Sigma c_i({n \atop i})$ and $c_i = \Delta^i f(i)$ where $(\Delta f)(i) = f(i+1) - f(i)$.

<u>Example 4.2</u>. In $\mathbb{V}$, $[{n \atop i}]$ is the number of i-dimensional subspaces of an n-dimensional space and if the field, K, of scalars has q elements then

$$[{n \atop i}] = \frac{k(n)+}{k(i)+k(n-i)+}$$

where

$$k(n) = q^n - 1$$

and

$$k(0)+ = k(0),\ k(i+1)+ = k(i+1)\cdot(k(i)+).$$

Not every strict $\mathbb{S}$-combinatorial function $f : E \to E$ is strict $\mathbb{V}$-combinatorial since the $[{n \atop i}]$ increase too rapidly.

The generalizations to several variables are straight-forward ($[{n,m \atop i,j}] = [{n \atop i}]\cdot[{n \atop j}]$, etc.)

Example 4.3. In $\mathbb{S}$, $x + y = \binom{x}{1}\cdot\binom{y}{0} + \binom{x}{0}\cdot\binom{y}{1}$, and $x\cdot y = \binom{x}{1}\binom{y}{1}$ are strict combinatorial and so are $(x+1)^y$ (but not x^y) and $x!$. In $\mathbb{V}$, $x + y$ not strict combinatorial and neither is the identity. However, the closure functor $\mathbb{V} : S \to \mathbb{V}$, of Example 2.5, is strict combinatorial.

Theorem 4.4. If $F : \mathbb{C}^1 \to \mathbb{C}^2$ is strict combinatorial then there exists a strict $\mathbb{S}$-combinatorial functor H such that

$$\begin{array}{ccc} C^1 & \xrightarrow{F} & C^2 \\ \uparrow V^1 & & \uparrow V^2 \\ \mathbb{S} & \dashrightarrow_{H} & \mathbb{S} \end{array}$$

commutes where V^i $(i = 1,2)$ is the appropriate closure functor.

Corollary 4.5. $[{n \atop i}]_{\mathbb{C}}$ is $\mathbb{S}$-combinatorial (as a functor of n).

Example.

$$[{n \atop i}]_{\mathbb{V}} = \frac{(q^n-1)\ldots(q^{n-i+1}-1)}{(q^i-1)\ldots(q-1)} = \sum_j c_j \binom{n}{j}$$

for some non-negative integers c_j.

5. Recursive equivalence. Now we turn to recursive equivalence. Two objects $\mathfrak{A}, \mathfrak{B}$ in a suitable category $\mathbb{C} = \mathbb{C}(\mathfrak{M})$ are said to be $\mathbb{C}$-recursively equivalent if there is an isomorphism $q : \mathfrak{A} \cong \mathfrak{B}$ in $\mathbb{C}(\mathfrak{M})$ which can be extended to a one-one partial recursive function. The equivalence classes under this relation are called $\mathbb{C}$-recursive equivalence types ($\mathbb{C}$-RETs, for short).

Finite $\mathbb{C}$-RETs may be identified with isomorphism types of objects in $\mathbb{C}$. We write $\langle \mathfrak{A} \rangle$ for the $\mathbb{C}$-RET of an object $\mathfrak{A}$.

A set A is said to be <u>effectively Dedekind finite</u> if there is no one-one partial recursive function which maps A onto a proper subset of itself. An object $\mathfrak{B} \in \mathbb{C}$ is said to be effectively Dedekind finite (Dedekind, for short) if its underlying set is effectively Dedekind finite. The $\mathbb{C}$-RETs of Dedekind objects are called <u>Dedekind $\mathbb{C}$-types</u>. The set of Dedekind $\mathbb{C}$-types is denoted $\Lambda(\mathbb{C})$. For each (classical) isomorphism type of infinite object in $\mathbb{C}$ there are continuum many $(c = 2^{\aleph_0})$ distinct Dedekind $\mathbb{C}$-types. This result is proved, as are most of the theorems below, by a Baire category argument and it turns out here that the representative of Dedekind $\mathbb{C}$-types contitute a second category set in a suitable Cantor space.

Recall that an <u>explicit index</u> of a finite set gives the cardinal of the set as well as its elements whereas a <u>standard index</u> merely allows a recursive enumeration of the set.

A combinatorial functor $F : \mathbb{C} \to \mathbb{C}'$ is said to be <u>finitary recursive</u> (<u>partial recursive</u>) if F maps ${}^{0}\mathbb{C}$ into the finite (the r.e.) algebraically closed objects of $\mathbb{C}'$ and (i) there is a recursive function f which assigns to each explicit index $\ulcorner \mathfrak{A} \urcorner$ of a finite object in ${}^{0}\mathbb{C}$ an explicit (a standard) index $\ulcorner F\mathfrak{A} \urcorner$ of $F\mathfrak{A}$ in $\mathbb{C}'$ and (ii) the map from $x \in \mathfrak{M}'$ to the explicit index $\ulcorner F^{\leftarrow} x \urcorner$ of $F^{\leftarrow} x$ in ${}^{0}\mathbb{C}$ is partial recursive.

So the recursive requirements are just effectivizations of the conditions in Section 2. Clearly finitary recursive combinatorial functors are partial recursive.

<u>Theorem 5.1</u>. (i) Partial recursive combinatorial functors F induce functions $\underline{F}$ on $\mathbb{C}$-RETs given by $\underline{F}(\underline{A}) = (F(\mathfrak{A}))$ where $A = \langle \mathfrak{A} \rangle$.

(ii) Partial recursive combinatorial functors are closed under composition. Similarly for finitary recursive ones. Moreover, with the notation of part (i), if $H = F \circ G$ then $\underline{H} = \underline{F} \circ \underline{G}$.

A pre-combinatorial operator G is said to be recursive if there is a recursive function which when applied to an explicit index of $\mathfrak{A}$ in ${}^{o}\mathbb{C}$ gives a standard index of $G(\mathfrak{A})$. Recursive precombinatorial operators induce partial recursive strict combinatorial functors. Also the closure functor V is finitary recursive, if the underlying functor is finitary recursive

A strict combinatorial functor is finitary recursive if and only if it induces a (recursive) number theoretic function with a recursive Stirling coefficient function.

6. Transfer principles. Finally we exhibit some of the transfer principles which stem from combinatorial functors. Throughout $\mathbb{C}, \mathbb{C}'$ will be assumed to be suitable categories.

Theorem 6.1. Let $F, G : \mathbb{C} \to \mathbb{C}'$ be finitary recursive combinatorial functors inducing functions $\underline{F}$, $\underline{G}$ on $\mathbb{C}$-RETs to $\mathbb{C}'$-RETs. Then the following are equivalent

(i) $\underline{F}(\underline{X}) = \underline{G}(\underline{X})$ for all $\mathbb{C}$-RETs $\underline{X}$ and

(ii) $\underline{F}(\underline{X}) = \underline{G}(\underline{X})$ for all Dedekind $\mathbb{C}$-RETs $\underline{X}$.

It is not true in general that if $\underline{F}(\underline{X}) = \underline{G}(\underline{X})$ for all finite $\mathbb{C}$-isomorphism types (= finite $\mathbb{C}$-RETs) that (i) and (ii) hold. Consider for example addition of order types then $\alpha + \beta = \beta + \alpha$ for all finite order types α, β but $\omega + 1 \neq 1 + \omega$ so equality cannot hold for all infinite $\mathbb{L}$-RETs. However, in the dimension case other properties do carry over.

Theorem 6.2. Hypotheses as for Theorem 6.1 plus $\mathbb{C}$ has dimension. Then (i), (ii) are equivalent to

(iii) $\underline{F}(\underline{X}) = \underline{G}(\underline{X})$ for all finite $\mathbb{C}$-RETs $\underline{X}$.

As a corollary all the usual associative, commutative and distributive laws concerning addition and multiplication and related laws for exponentiation for cardinals carry over to $-RETs. We have also shown that provided the isomorphisms are <u>uniform</u> then identities do carry over from finite types to infinite ones. The full technical details may be found in [Crossley and Nerode 1974].

For the case where $\underline{F}(\underline{X}) = \underline{G}(\underline{X})$ holds only sometimes, we have to consider chains of objects. A <u>chain</u> is a sequence

$$\mathfrak{A}^{o} \xrightarrow{p^1} \mathfrak{A}^{1} \xrightarrow{p^2} \cdots$$

where the p^i are $\mathbb{C}$-morphisms and the $\mathfrak{A}^i$ are objects of $\mathbb{C}$. The <u>length</u> of the chain is the number of morphisms. Two chains $\mathfrak{A}^{o} \xrightarrow{p^1} \cdots$ and $\mathfrak{L}^{o} \xrightarrow{q^1} \cdots$ of length n are said to be n-equivalent if there are $\mathbb{C}$-isomorphisms $r^o, r^1, \ldots, r^n$ such that

$$\begin{array}{ccccccc}
\mathfrak{A}^{o} & \xrightarrow{p^1} & \mathfrak{A}^{1} & \xrightarrow{p^2} & \cdots & \xrightarrow{p^n} & \mathfrak{A}^{n} \\
\Big| r^o & & \Big| r^1 & & \cdots & & \Big| r^n \\
\mathfrak{L}^{o} & \xrightarrow[q^1]{} & \mathfrak{L}^{1} & \xrightarrow[q^2]{} & \cdots & \xrightarrow[q^n]{} & \mathfrak{L}^{n}
\end{array}$$

commutes.

The equivalence classes are called n-<u>chain</u> <u>types</u>.

A <u>frame</u> is a subset of ${}^{o}\mathbb{C}$ closed under intersection. An object $\mathfrak{A}$ is <u>attainable</u> from a frame F if for every finite set $B \subseteq \mathfrak{A}$ there exists $\mathfrak{L} \in F$ such that $B \subseteq \mathfrak{L} \subseteq F$.
For a frame F if $B \subseteq \mathfrak{L}$ for some $\mathfrak{L} \in F$ then there is a smallest $\mathfrak{L} = C_F(B)$ with this property. A frame F is <u>recursive</u> if
(i) the set of indices of elements of F is r.e. and
(ii) the number-theoretic function with domain $\{\ulcorner B \urcorner : C_F(B)$ is defined$\}$ (where $\ulcorner B \urcorner$ is an explicit index of B) given by $\ulcorner B \urcorner \leadsto \ulcorner C_F(B) \urcorner$, is partial recursive. If R is a relation on

on chain types then a frame F is an ^{n}R-frame if every n-chain whose objects are in F is in R. (Note that we are using $\mathbb{C}\times\ldots\times\mathbb{C}$, m factors, instead of C if R is an m-ary relation.)

Definition 6.3. The extension $\mathbb{C}(^{n}R)$ of a relation R on chain types is the set of all Dedekind $\mathbb{C}$-types $\underline{X}$ such that some representative $\mathcal{X} \in \underline{X}$ is attainable from a recursive ^{n}R-frame. We write $\mathbb{C}(R)$ for $\mathbb{C}(^{o}R)$ noting that 0-chain types are just isomorphism types. Now finitary recursive combinatorial functors, F say, induce functions $\underline{F}$ on chain types and we have:

Theorem 6.4. Let $F,G : \mathbb{C} \to \mathbb{C}'$ be finitary recursive combinatorial functors and let ^{n}R be the set of finite n-chain type solutions $\underline{C}$ to $\underline{F}(\underline{C}) = \underline{G}(\underline{C})$ then $\mathbb{C}(^{n}R) \supseteq \{\underline{X} \in \Lambda(\mathbb{C}) : \underline{F}(\underline{X}) = \underline{G}(\underline{X})\}$. If $\mathbb{C} = \mathbb{C}(\mathfrak{M})$ has dimension and $\mathfrak{M}$ has degree 1 then the other inclusion also holds.

$\mathfrak{M}$ has degree 1 if whenever $a \in cl\{a_o,\ldots,a_{n-1}\}$ there is a formula ϕ such that $\mathfrak{M} \vdash \phi[a_o,\ldots,a_{n-1},a]$ and $\mathfrak{M} \vdash \exists! v_n \phi(v_o,\ldots,v_n)\ [a_o,\ldots,a_{n-1}]$.

Our final theorem holds when all the categories have dimension. We state it only for the case of relations on a single category for simplicity. Let $X_\omega E$ denote the set of those sequences of natural numbers such that the sequence is zero for all but finitely many places. We identify an n-ary relation R on $X^n E = E\times\cdots\times E$, n factors, with the relation R' on $X_\omega E$ such that for $x \in X_\omega E$ we have $x \in R'$ if, and only if $x|n \in R$ where $x|n$ is the first n elements of the sequence x. A relation R' which arises in this way is said to be a finitary relation. Then the notion of $\mathbb{C}(R)$ extends naturally to give $\mathbb{C}(R')$.

Theorem 6.5. (Compactness theorem for categories with dimension.) Let $\mathcal{R}$ be a set of r.e. finitary relations on $X_\omega E$. If every finite subset of $\mathcal{R}$ is satisfied by some element of $X_\omega E$ then there is a sequence of Dedekind $\mathbb{C}$-types which simultaneously satisfies every extension, $\mathbb{C}(R)$, of a relation R in $\mathcal{R}$.

Monash University
Melbourne, Australia

Cornell University
Ithaca, N.Y.,
U.S.A.

BIBLIOGRAPHY

Aczel, P.H.G. [1966] D. Phil thesis Oxford (1966).

Crossley, J.N. [1969] Constructive Order Types (monograph) Amsterdam

Crossley, J.N. [1970] Recursive Equivalence, Bull. L.M.S. 2 (1970) 129-151.

Crossley, J.N. & A. Nerode [1974] Combinatorial Functors (Ergebnisse der Mathematik und ihrer Grenzgebiete Bd. 81, Springer).

Dekker, J.C.E. [1966] Les functions combinatores et les isols, Paris.

Dekker, J.C.E. & J. Myhill [1960] Recursive Equivalence Types, University of California publications in mathematics, n.s. 3 (1960) 67-214.

Ellentuck, E. [1965] The universal properties of Dedekind finite cardinals, Ann. of Math., 82 (1965) 225-248.

EQUALITY BETWEEN FUNCTIONALS

Harvey Friedman*

The λ-calculus, in both its typed and untyped forms, has primarily been regarded as an attempt to formalize the concept of _rule_ or _process_, and ultimately to provide a new foundation for mathematics. It seems fair to say that this aspect of the λ-calculus is currently in an embryonic state of development, awaiting further conceptual advances.

There is, however, another aspect of the _typed_ λ-calculus, which is readily understood. This is its connection with the full classical finite type structure over ω, (i.e., with the functionals of finite type). In an obvious way, each closed term in the typed λ-calculus defines a functional of finite type over ω. Call a functional _simple_ if it is given by some closed term in the typed λ-calculus.

Several definitions of _convertibility_ between terms in the λ-calculus, with or without types, have been considered. The motivation for introducing these definitions has primarily been to analyze the notion of the _identity_ between rules, or processes. The relation $\vdash s = t$ defined in the text, is equivalent to one of these definitions of convertibility. It is easy to see that any two convertible terms define the same functional of finite type. We show here that any two non-convertible terms define different functionals of finite type.[1] This, coupled with the known decidability of convertibility, tells us that "equality between simple functionals is recursive."

Let us call two functionals _strongly unequal_ if they differ everywhere. We show that, in contrast to the above, "strong inequality between simple functionals is as complicated as the set of true sentences of type theory over ω."

Augment the typed λ-calculus by (primitive) recursion operators, and call the result the _R-λ-calculus_. The functionals denoted by closed R-terms are the primitive recursive functionals of finite type. We conclude the paper by demonstrating that "equality between primitive recursive functionals is complete Π_1^1."

---

*This research was partially supported by NSF PO38823.

[1] This result was obtained in 1970.

1. The typed λ-calculus.

We first describe the syntax of the typed λ-calculus.

The type symbols are given by i) o is a type symbol ii) (σ,τ) is a type symbol if σ,τ are. The variables are written x_n^σ. The terms s , their types, their sets of free variables FV(s) , and their sets of bound variables BV(s) are given by i) x_n^σ is a term of type σ , $FV(x_n^\sigma) = \{x_n^\sigma\}$, $BV(x_n^\sigma) = \phi$ ii) if s is a term of type (σ,τ) , t a term of type σ , then (st) is a term of type τ , $FV((st)) = FV(s) \cup FV(t)$, $BV((st)) = BV(s) \cup BV(t)$ iii) if s is a term of type τ , y a variable of type σ , then (λys) is a term of type (σ,τ) , $FV((\lambda ys)) =$ $= FV(s) - \{y\}$, $BV((\lambda ys)) = BV(s) \cup \{y\}$. The collection of all terms is denoted by Tm .

We now describe the semantics we will use for the typed λ-calculus.

A pre-structure is a system $(\{D^\sigma\}, \{A_{\sigma\tau}\})$, where D^σ is a nonempty set, for each type symbol σ , and $A_{\sigma\tau}: D^{(\sigma,\tau)} \times D^\sigma \to D^\tau$, for each type symbols σ,τ. We require the following extensionality condition: if $x,y \in D^{(\sigma,\tau)}$ and $(\forall z \in D^\sigma)(A_{\sigma\tau}(x,z) = A_{\sigma\tau}(y,z))$ then $x = y$.

An assignment in the system $(\{D^\sigma\},\{A_{\sigma\tau}\})$ is a function f whose domain is the set of all variables, and such that $f(x_n^\sigma) \in D^\sigma$. The set of all assignments is denoted Asg. If y is a variable then f_α^y is given by $f_\alpha^y(x) = f(x)$ for $y \neq x$; $f_\alpha^y(y) = \alpha$.

A structure is a system $(\{D^\sigma\},\{A_{\sigma\tau}\}, Val)$ such that $(\{D^\sigma\},\{A_{\sigma\tau}\})$ is a pre-structure, and $Val: Tm \times Asg \to \bigcup_\sigma D^\sigma$, such that the following clauses hold: i) $Val(x_n^\sigma,f) = f(x_n^\sigma)$ ii) $Val((st),f) = A_{\sigma\tau}(Val(s,f),Val(t,f))$, where s has type (σ,τ) , t has type σ iii) for all $\alpha \in D^\sigma$, $A_{\sigma\tau}(Val((\lambda xs),f),\alpha) = Val(s,f_\alpha^x)$, where s is of type τ , x is of type σ .

Suppose $(\{D^\sigma\},\{A_{\sigma\tau}\})$ is a pre-structure. Then there is at most one function Val such that $(\{D^\sigma\},\{A_{\sigma\tau}\},Val)$ is a structure. Thus we sometimes refer to structures $(\{D^\sigma\},\{A_{\sigma\tau}\})$, meaning that for some Val, $(\{D^\sigma\},\{A_{\sigma\tau}\},Val)$ is a structure.

For any structure $(\{D^\sigma\},\{A_{\sigma\tau}\})$, we write $(\{D^\sigma\},\{A_{\sigma\tau}\}) \models s = t[f]$, just in case $Val(s,f) = Val(t,f)$, (where $(\{D^\sigma\},\{A_{\sigma\tau}\},Val)$ is a structure). Write

$(\{D^\sigma\},\{A_{\sigma\tau}\}) \models s = t$ if and only if $(\{D^\sigma\},\{A_{\sigma\tau}\}) \models s = t[f]$ for all assignments f. Below, we will often leave off the subscripts of $A_{\sigma\tau}$.

Let B be a nonempty set. We introduce the important example of a structure, T_B, often referred to as the full type structure over B. $T_B = (\{B^\sigma\},\{A_{\sigma\tau}\})$, where $B^0 = B$, $B^{(\sigma,\tau)} = (B^\tau)^{B^\sigma}$, $A_{\sigma\tau}(x,y) = x(y)$.

We wish to show that, for all infinite B, the relation $T_B \models s = t$ is decidable, and is the same for all infinite B. As an intermediate step, we establish a <u>completeness theorem</u> for the typed λ-calculus.

If s is a term, x a variable, t a term of the same type as x, then let s^x_t denote the substitution of the term t for each free occurrence of x in s. This may be inductively defined by i) $x^x_t = t$ ii) $y^x_t = y$ for variables $y \neq x$ iii) $(rs)^x_t = (r^x_t s^x_t)$ iv) $(\lambda xs)^x_t = (\lambda xs)$ v) $(\lambda ys)^x_t = (\lambda y s^x_t)$, for variables $y \neq x$. A <u>substitution</u> is a function g from all variables into terms, such that $g(x)$ has the same type as x. Similarly, let $s(g)$ denote the simultaneous substitution of each free occurrence of each variable y in s, by $g(y)$.

We now introduce <u>axioms</u> and <u>rules</u> to the typed λ-calculus.

1. $(\lambda xs) = (\lambda y s^x_y)$, if $y \notin FV(s) \cup BV(s)$.
2. $((\lambda xs)t) = s^x_t$, if $BV(s) \cap FV(t) = \phi$.
3. $(\lambda x(sx)) = s$, if $x \notin FV(s)$.
4. $s = s$, $\dfrac{s=t}{t=s}$, $\dfrac{s=t,\ t=r}{s=r}$.
5. $\dfrac{s=t}{(rs)=(rt)}$, $\dfrac{s=t}{(sr)=(tr)}$, $\dfrac{s=t}{(\lambda xs)=(\lambda xt)}$.

We first prove the soundness theorem. Fix a structure $M = (\{D^\sigma\},\{A_{\sigma\tau}\},Val)$.

LEMMA 1. $Val(s^x_t,f) = Val(s,f^x_{Val(t,f)})$, if $BV(s) \cap FV(t) = \phi$.

<u>Proof</u>: Fix t, and use induction on s.

LEMMA 2. $Val(s^x_y,f^y_\alpha) = Val(s,f^x_\alpha)$, if $y \notin FV(s) \cup BV(s)$.

<u>Proof</u>: From Lemma 1.

LEMMA 3. $M \models (\lambda xs) = (\lambda y s^x_y)$, if $y \notin FV(s) \cup BV(s)$.

Proof: $A(\mathrm{Val}((\lambda xs),f),\alpha) = \mathrm{Val}(s,f^x_\alpha) = \mathrm{Val}(s^x_y,f^x_\alpha) = A(\mathrm{Val}((\lambda ys^x_y),f),\alpha)$, by Lemma 2.

LEMMA 4. $\mathrm{Val}(((\lambda xs)t),f) = \mathrm{Val}(s^x_t,f)$, if $BV(s) \cap FV(t) = \phi$.

Proof: $\mathrm{Val}(((\lambda xs)t),f) = A(\mathrm{Val}((\lambda xs),f), \mathrm{Val}(t,f)) = \mathrm{Val}(s,f^x_{\mathrm{Val}(t,f)}) =$
$= \mathrm{Val}(s^x_t,f)$, by Lemma 1.

LEMMA 5. $\mathrm{Val}((\lambda x(sx)),f) = \mathrm{Val}(s,f)$, if $x \notin FV(s)$.

Proof: $A(\mathrm{Val}((\lambda x(sx)),f),\alpha) = \mathrm{Val}((sx),f^x_\alpha) = A(\mathrm{Val}(s,f^x_\alpha), \mathrm{Val}(x,f^x_\alpha)) =$
$A(\mathrm{Val}(s,f),\alpha)$.

LEMMA 6. If $M \models s = t$ then $M \models (\lambda xs) = (\lambda xt)$.

Proof: Fix f . Then $A(\mathrm{Val}((\lambda xs),f),\alpha) = \mathrm{Val}(s,f^x_\alpha) = \mathrm{Val}(t,f^x_\alpha) =$
$= A(\mathrm{Val}((\lambda xt),f),\alpha)$.

THEOREM 1. (Soundness). If $\vdash s = t$, then every structure $M \models s = t$.

Proof: By induction on the proof of $s = t$, using Lemmas 1-6.

For the Proof of completeness, we consider a particular structure defined from the relation $\vdash$. Let $[s]$, for terms s , be $\{t: \vdash s = t\}$. It is clear that the $[s]$ are the equivalence classes of the equivalence relation $\vdash s = t$. This is because of the rules 4.

We wish to define a specific $M_0 = (\{D^\sigma\},\{A_{\sigma\tau}\})$. Take $D^\sigma = \{[s]: s$ is of type $\sigma\}$. Define $A_{\sigma\tau}([s],[t]) = [(st)]$, where s is of type (σ,τ) , t is of type τ .

We must now check that M_0 is well-defined. Firstly we remark that if $\vdash s = t$, then s and t are of the same type. Secondly, note that if $\vdash s = s'$, $\vdash t = t'$, then $\vdash (st) = (s't')$, by rules 5.

LEMMA 7. M_0 is a pre-structure.

Proof: Let $[s],[t] \in D^{(\sigma,\tau)}$. Suppose that for all $r \in D^\sigma$, we have

$[(sr)] = [(tr)]$. Then let x be a variable not free in either s or t . We have $[(sx)] = [tx]$. Hence $\vdash (sx) = (tx)$. By rule 5, $\vdash (\lambda x(sx)) = (\lambda x(tx))$. By axiom 3, $\vdash (\lambda x(sx)) = s$, $\vdash (\lambda x(tx)) = t$. By rules 4, $\vdash s = t$. Hence $[s] = [t]$.

Let us call a substitution g , <u>regular</u>, just in case for all variables x,y , $FV(g(x)) \cap BV(g(y)) = \phi$.

LEMMA 8. Let b be a finite set of variables, s a term. Then there is a term t such that $\vdash s = t$, $FV(s) = FV(t)$, and $BV(t) \cap b = \phi$.

<u>Proof</u>: By successive applications of axiom 1 and rules 4.

Let f be an assignment for M_0 , s a term. We wish to define $Val(s,f)$. By Lemma 8, let g be a regular substitution such that each $f(x) = [g(x)]$. Choose t to be a term such that $\vdash s = t$, and $BV(t) \cap FV(f(x)) = \phi$, for all $x \in FV(t)$, again by Lemma 8. Set $Val(s,f) = [t(g)]$. We must now show that $Val(s,f)$ is well-defined.

LEMMA 9. Suppose g_1, g_2 are regular substitutions such that $\vdash g_1(x) = g_2(x)$, for all variables x . Suppose $\vdash s = t$, $BV(s) \cap FV(g_1(x)) = BV(t) \cap FV(g_2(y)) = \phi$, for all $x \in FV(s)$, $y \in FV(t)$. Then $\vdash s(g_1) = t(g_2)$.

<u>Proof</u>: By induction on the cardinality k of $FV(s) \cup FV(t)$. The case $k = 0$ is trivial. Let $k = n + 1$, and assume true for n . Let $x \in FV(s) \cup FV(t)$. Choose w of the same type as x , so that $w \notin FV(s) \cup FV(t) \cup BV(s) \cup BV(t)$, $w \notin FV(g_1(y)) \cup FV(g_2(z))$, for all $y \in FV(s)$, $z \in FV(t)$. We have $\vdash (\lambda xs) = (\lambda xt)$, and $\vdash (\lambda xs) = (\lambda ws^x_w)$, $(\lambda xt) = (\lambda wt^x_w)$. So $\vdash (\lambda ws^x_w) = (\lambda wt^x_w)$. By induction hypothesis, $\vdash (\lambda ws^x_w)(g_1) = (\lambda wt^x_w)(g_2)$. Let $h_1 = (g_1)^x_x$, $h_2 = (g_2)^x_x$. Then $\vdash (\lambda w(s(h_1))^x_w) = (\lambda w(t(h_2))^x_w)$. Hence $\vdash (\lambda w(s(h_1))^x_w) g_1(x) = (\lambda w(t(h_2))^x_w) g_2(x)$. So $\vdash (s(h_1))^x_{g_1(x)} = (t(h_2))^x_{g_2(x)}$. Hence $\vdash s(g_1) = t(g_2)$.

LEMMA 10. $Val(s,f)$ is well defined, and $Val(s,f) = Val(t,f)$ if $\vdash s = t$.

<u>Proof</u>: Obvious from Lemma 9.

We now wish to show that M_0 is a structure. Write $f = [g]$, for M_0-assignments f , if g is a regular substitution and $f(x) = [g(x)]$.

LEMMA 11. $Val(x,f) = f(x)$, for variables x .

<u>Proof</u>: Let $f = [g]$. Then $Val(x,f) = [x(g)] = [g(x)] = f(x)$.

LEMMA 12. $Val((st),f) = A(Val(s,f),Val(t,f))$.

<u>Proof</u>: Let $f = [g]$. Choose s',t' so that $\vdash s = s'$, $\vdash t = t'$, $BV(s') \cap FV(g(x)) = BV(t') \cap FV(g(x)) = \phi$, for all $x \in FV(s') \cup FV(t')$. Then $Val((st),f) = [(s't')(g)] = [(s'(g)t'(g))] = A([s'(g)],[t'(g)]) = A(Val(s,f),$ $Val(t,f))$.

LEMMA 13. $A(Val((\lambda xs),f),[t]) = Val(s,f^x_{[t]})$.

<u>Proof</u>: Choose $f^x_{[t]} = [g]$. Let $\vdash s = s'$, $BV(s') \cap FV(g(y)) = \phi$, for all $y \in FV(s')$. Then $Val((\lambda xs),f) = [(\lambda xs')(g)]$. Let $h = g^x_x$. Then $(\lambda xs')(g) =$ $= (\lambda xs'(h))$. Note that $\vdash (\lambda xs'(h))(t) = (\lambda xs'(h))(g(x)) = s'(h)^x_{g(x)} = s'(g)$. Hence $A(Val((\lambda xs),f),[t]) = [s'(g)] = Val(s,f^x_{[t]})$.

LEMMA 14. $M_0 = (\{D^\sigma\},\{A_{\sigma\tau}\}, Val)$ is a structure.

<u>Proof</u>: By lemmas 11-13.

THEOREM 2. (Completeness). Let s,t be terms. The following are equivalent: i) $\vdash s = t$ ii) for all structures M , $M \models s = t$ iii) $M_0 \models s = t$.

<u>Proof</u>: By Theorem 1, we must simply show that $M_0 \models s = t$ implies $\vdash s = t$. Suppose $M_0 \models s = t$. Let $f = [g]$, where g is the identity map, and choose s',t' such that $\vdash s = s'$, $\vdash t = t'$, and $BV(s') \cap FV(s') = BV(t') \cap FV(t') = \phi$. Then $Val(s,f) = [s']$, $Val(t,f) = [t']$. Hence $[s'] = [t']$, and so $\vdash s' = t'$. Hence $\vdash s = t$, and we are done.

Let $M = (\{D^\sigma\},\{A_{\sigma\tau}\},Val_1)$, $N = (\{E^\sigma\},\{B_{\sigma\tau}\},Val_2)$ be structures. A system $\{f_\sigma\}$ is called a <u>partial</u> <u>homomorphism</u> from M onto N just in case i) each f_σ is a partial surjective map from D_σ onto E_σ ii) $f_{\sigma\tau}(x)$ is the unique element of $E^{(\sigma,\tau)}$ (if it exists) such that $f_\tau(A(x,y)) = B(f_{\sigma\tau}(x), f_\sigma(y))$, for all $y \in Dom(f_\sigma)$. Note that $\{f_\sigma\}$ is determined by f_0 , and this definition does not involve Val. The following Lemma does.

LEMMA 15. If $\{f_\sigma\}$ is a partial homomorphism from M onto N , and g is an M-assignment, h is an N-assignment, $f_\sigma(g(x)) = h(x)$ for variables x of type σ , then $f_\sigma(Val_1(s,g)) = Val_2(s,h)$, for terms s of type σ .

<u>Proof</u>: By induction on s , where $M,N,\{f_\sigma\}$ are fixed. $f_\sigma(Val_1(x,g)) = $
$= f_\sigma(g(x)) = h(x) = Val_2(x,h)$, for variables x of type σ .

Now $f_\tau(Val_1((st),g)) = f_\tau(A(Val_1(s,g),Val_1(t,g)))$. By induction hypothesis, $f_{\sigma\tau}(Val_1(s,g)) = Val_2(s,h)$, $f_\sigma(Val_1(t,g)) = Val_2(t,h)$. Hence
$f_\tau(A(Val_1(s,g)),Val_1(t,g)) = B(Val_2(s,h),Val_2(t,h)) = Val_2((st),h)$.

Finally, we must show $f_{\sigma\tau}(Val_1((\lambda xs),g)) = Val_2((\lambda xs),h)$. To do this, let $y \in Dom(f_\sigma)$. We must show $f_\tau(A(Val_1((\lambda xs),g),y)) = B(Val_2((\lambda xs),h),f_\sigma(y))$. Now $f_\tau(A(Val_1((\lambda xs),g),y)) = f_\tau(Val_1(s,g^x_y)) = Val_2(s,h^x_{f_\sigma(y)}) = B(Val_2((\lambda xs),h),f_\sigma(y))$. We are done.

LEMMA 16. Suppose there is a partial homomorphism from M onto N . Then $M \models s = t$ implies $N \models s = t$, for any terms s,t .

<u>Proof</u>: Let $\{f_\sigma\}$ be a partial homomorphism, and assume $M \models s = t$. Let h be an N-assignment. Choose an M-assignment g so that $h(x) = f_\sigma(g(x))$, for variables x of type σ . Then $Val_2(s,h) = f_\sigma(Val_1(s,g)) = f_\sigma(Val_1(t,g)) = Val_2(t,h)$, by Lemma 15, for terms s,t of type σ . Hence $N \models s = t$.

For sets B, let $|B|$ be the cardinal of B .

LEMMA 17. Let $M = (\{D^\sigma\},\{A_{\sigma\tau}\},Val)$ be a structure $|D^0| \le |B|$. Then there is a partial homomorphism from T_B onto M .

<u>Proof</u>: We define $\{f_\sigma\}$ by induction on the type symbol σ . Let f_0 be any partial surjective map from B onto D^0 . Suppose f_σ, f_τ have been defined,

surjectively, according to the clauses for being a partial homomorphism. Define $f_{\sigma\tau}(x)$ to be the unique element of $D^{(\sigma,\tau)}$ (if it exists) such that $f_{\tau}(x(y)) = A(f_{\sigma\tau}(x), f_{\sigma}(y))$, for all $y \in Dom(f_{\sigma})$. We must show that $f_{\sigma\tau}$ is surjective. Let $z \in D_{(\sigma,\tau)}$. Choose $x \in B^{(\sigma,\tau)}$ so that for all $y \in Dom(f_{\sigma})$, $x(y) \in f_{\tau}^{-1}(A(z, f_{\sigma}(y)))$. Then $f_{\sigma\tau}(x) = z$.

THEOREM 3. (Extended Completeness). Let s,t be terms, B an infinite set. The following are equivalent: i) $\vdash s = t$ ii) for all structures M, $M \models s = t$ iii) $T_B \models s = t$.

<u>Proof</u>: By Theorem 2, it suffices to show that $T_B \models s = t$ implies $M_0 \models s = t$. By Lemma 17, there is a partial homomorphism from T_B onto M. By Lemma 16, if $T_B \models s = t$ then $M \models s = t$.

LEMMA 18. The relation $\vdash s = t$ is recursive.

<u>Proof</u>: This follows from the following known fact about the typed λ-calculus (even with recursion operators): every term reduces to a unique irreducible term, up to changes in bound variables, no matter how the reductions are performed (see Sanchis [2], Tait [3], and Barendregt [1] for elaboration).

COROLLARY. If B is infinite then the relation $T_B \models s = t$ is recursive, and is independent of the size of B.

Let B be finite, $g: B \to B$. Define $g^1 = g$, $g^{k+1} = g \circ g^k$. We will show that the extended completeness theorem fails for B.

LEMMA 19. There are $i(g), j(g) \geq 1$ such that for $n > i$, we have $g^n = g^{n+j}$.

Fix x to be a variable of type $(0,0)$. Let $x^1 = x$, $x^{k+1} = x \circ x^k = (\lambda y(x(x^k y)))$.

THEOREM 4. For each nonempty finite B, there are terms s,t of type $(0,0)$ with $T_B \models s = t$, such that not $\vdash s = t$.

<u>Proof</u>: For each $g: B \to B$ define $i(g), j(g)$ as in Lemma 19. Choose i

greater than each $i(g)$, and set $j = \prod_g j(g)$. Then for each $g: B \to B$ we have $g^i = g^{i+j}$. Hence $T_B \models x^i = x^{i+j}$. To see that not $\vdash s = t$, consider T_ω . Note that not $T_\omega \models x^i = x^{i+j}$, since x may be interpreted as the successor function on ω . We are done.

Now let $M \models s \neq t$ mean not $M \models s = t[f]$, for all M-assignments f . (We will often write $M = s \neq t[f]$ for not $M \models s = t[f]$) . Does the Corollary to Theorem 3 hold for the relation $T_B \models s \neq t$? Below, we give a negative answer.

We introduce a many-sorted predicate calculus (with equality), $\mathcal{L}$, appropriate for the theory of functionals of finite type over a nonempty domain. Specifically, the <u>atomic</u> <u>formulae</u> of $\mathcal{L}$ are written $s = t$, for terms s,t , of the typed λ-calculus of the same type. The <u>formulae</u> of $\mathcal{L}$ are obtained from the atomic formulae by using $\sim,\&,\forall$. The $\forall$-quantifiers quantify over a given type only. Thus $M \models \varphi[f]$, for structures M , formulae φ of $\mathcal{L}$, and M-assignments f , is defined in the obvious way. $\exists, \vee$ are introduced as abbreviations in the standard manner. Take $M \models \varphi$ to mean $M \models \varphi[f]$ for all M-assignments f .

A formula φ of $\mathcal{L}$ is called <u>existential</u> if it is of the form $(\exists x)(s = t)$.

Let 0 be the closed term $(\lambda y(\lambda xy))$, 1 the closed term $(\lambda y(\lambda xx))$, where x,y are distinct variables of type 0 . For terms s,t , let $\langle s,t\rangle$ be the term $(\lambda x((xs)t))$, where $x \notin FV(s) \cup FV(t)$, so that $\langle s,t\rangle$ has type 0 .

LEMMA 20. If B has at least two elements, then $T_B \models 0 \neq 1$.

LEMMA 21. If s,u have the same type, t,v have the same type, then $T_B \models \langle s,t\rangle = \langle u,v\rangle \to (s = u \;\&\; t = v)$.

LEMMA 22. If φ is existential, there is an existential ψ with the same free variables, such that $T_B \models \psi \leftrightarrow {\sim}\varphi$, for all B with at least two elements.

<u>Proof</u>: Let φ be $(\exists x)(s = t)$. Note that by Lemma 20, $T_B \models (\exists y)((ys) = {}= 0 \;\&\; (yt) = 1) \leftrightarrow s \neq t$, where $y \notin FV(s) \cup FV(t)$, $y \neq x$. By Lemma 21, $T_B \models (\exists y)\ (\langle (ys),(yt)\rangle = \langle 0,1\rangle) \leftrightarrow s \neq t$. Hence $T_B \models (\forall x)(\exists y)\ (\langle (ys),(yt)\rangle = {}= \langle 0,1\rangle) \leftrightarrow (\forall x)(s \neq t)$. So $T_B \models (\exists z)(\forall x)(\langle ((zx)s),((zx)t)\rangle = {}= \langle 0,1\rangle) \leftrightarrow (\forall x)(s \neq t)$, where $z \notin FV(s) \cup FV(t)$, $z \neq x,y$. Hence $T_B \models (\exists z)$

$((\lambda x\langle((zx)s),((zx)t)\rangle) = (\lambda x\langle 0,1\rangle)) \leftrightarrow (\forall x)(s \neq t)$.

LEMMA 23. If φ,ψ are existential, then there is an existential ρ with the same free variables as $\varphi \& \psi$, such that $T_B \models \rho \leftrightarrow (\varphi \& \psi)$, for all B with at least two elements.

Proof: Let φ be $(\exists x)(s = t)$, ψ be $(\exists x)(u = v)$, (where φ,ψ may have had their bound variable changed to x). Then $T_B \models (\exists x)(\langle s,u\rangle = \langle t,v\rangle) \leftrightarrow$ $\leftrightarrow ((\exists x)(s = t) \& (\exists x)(u = v))$.

LEMMA 24. If φ is existential, then there is an existential ρ with the same free variables as $(\exists x)(\varphi)$, such that $T_B \models \rho \leftrightarrow (\exists x)(\varphi)$, for all B with at least two elements.

Proof: Let φ be $(\exists y)(s = t)$, where $y \neq x$, (where φ may have had its bound variable changed). Then $T_B \models (\exists z)(s^{x\ \ \ \ y}_{(z0)(z1)} = t^{x\ \ \ \ y}_{(z0)(z1)}) \leftrightarrow (\exists x)(\exists y)(s = t)$.

LEMMA 25. For each formula φ of $\mathcal{L}$, we can effectively find an existential ψ with the same free variables, such that $T_B \models \varphi \leftrightarrow \psi$, for each B with at least two elements.

Proof: From Lemmas 22, 23, 24.

LEMMA 26. There is a one-one total recursive function f such that for each formula φ of $\mathcal{L}$, $f(\varphi)$ is an existential formula with the same free variables as φ, and $T_B \models \varphi \leftrightarrow \psi$, for each B with at least two elements.

Proof: This is an effective version of Lemma 25, obtained from corresponding effective versions of Lemmas 22, 23, 24.

THEOREM 5. For each B, the set of sentences φ of $\mathcal{L}$ such that $T_B \models \varphi$, is one-one reducible to the relation $T_B \models s \neq t$.

Proof: We can assume that B has at least two elements (or for that matter, is infinite), since otherwise $\{\varphi: T_B \models \varphi\}$ is recursive. Note that for sentences φ of $\mathcal{L}$, $T_B \models \varphi$ if and only if not $T_B \models (\sim\varphi)$ if and only if not $T_B \models f((\sim\varphi))$ if and only if $T_B \models s \neq t$, where $f((\sim\varphi)) = (\exists x)(s = t)$.

2. The typed λ-calculus with primitive recursion.

We will refer to this extension of the typed λ-calculus as the R-λ-calculus. The R-λ-calculus has the additional symbols $0, N$, and R_σ, for each type symbol σ. The variables of the R-λ-calculus are the same as the variables of the variables of the typed λ-calculus.

The terms s, their types, their sets of free variables $FV(s)$, and their sets of bound variables $BV(s)$ are given by i) x_n^σ is a term of type σ, $FV(x_n^\sigma) = \{x_n^\sigma\}$, $BV(x_n^\sigma) = \emptyset$ ii) if s is a term of type (σ,τ), t a term of type σ, then (st) is a term of type τ, $FV((st)) = FV(s) \cup FV(t)$, $BV((st)) = BV(s) \cup BV(t)$ iii) if s is a term of type τ, y a variable of type σ, then (λys) is a term of type (σ,τ), $FV((\lambda ys)) = FV(s) - \{y\}$, $BV((\lambda ys)) = BV(s) \cup \{y\}$ iv) 0 is a term of type 0, $FV(0) = \emptyset$, $BV(0) = \emptyset$ v) N is a term of type $(0,0)$, $FV(N) = \emptyset$, $BV(N) = \emptyset$ vi) R_σ is a term of type $((\sigma,(0,\sigma)),(\sigma,(0,\sigma)))$, $FV(R_\sigma) = \emptyset$, $BV(R_\sigma) = \emptyset$.

We let $(s_1,\ldots,s_{n+1})$ be $((s_1 s_2),\ldots,s_{n+1})$.

Let $(\{D^\sigma\},\{A_{\sigma\tau}\})$ be a pre-structure. It will be convenient to assume that the D^σ are disjoint. Let $A(x_1,\ldots,x_{n+1})$, for appropriate $x_1,\ldots,x_{n+1}$ be $A(A(x_1,x_2),\ldots,x_{n+1})$, where each occurrence of A denotes the appropriate $A_{\sigma\tau}$.

A system $(\{D^\sigma\},\{A_{\sigma\tau}\},Val)$ is an R-structure just in case i) $(\{D^\sigma\},\{A_{\sigma\tau}\})$ is a pre-structure ii) $D^0 = \omega$ iii) $Val(x_n^\sigma,f) = f(x_n^\sigma)$ iv) $Val((st,f) = A(Val(s,f),Val(t,f))$ v) for all $\alpha \in D^\sigma$, $A(Val((\lambda xs),f),\alpha) = Val(s,f_\alpha^x)$, where s is of type σ vi) $D^0 = \omega$ vii) $Val(0,f) = 0$ viii) $A(Val(N,f),k) = k+1$ ix) $A(Val(R_\sigma,f),y,z,0) = z$, $A(Val(R_\sigma,f),y,z,k+1) = A(y,A(Val(R_\sigma,f),y,z,k),k)$ for $y \in D^{(\sigma,(0,\sigma))}$, $z \in D^\sigma$, $k \in \omega$.

Note that if $(\{D^\sigma\},\{A_{\sigma\tau}\})$ is a pre-structure, there is at most one Val such that $(\{D^\sigma\},\{A_{\sigma\tau}\}, Val)$ is an R-structure. Thus we may refer to the R-structure $(\{D^\sigma\},\{A_{\sigma\tau}\})$.

Obviously, we may view T_ω as an R-structure just as we viewed T_ω as a structure in section 1.

As in section 1, we write $M \models s = t[f]$ to mean $Val(s,f) = Val(t,f)$, and $M \models s = t$ to mean $Val(s,f) = Val(t,f)$, for all assignments f. In this section we will show that the relation $T_\omega \models s = t$ is complete Π^1_1.

Let $M = (\{D_\sigma\},\{A_{\sigma\tau}\},Val_1)$, $N = (\{E_\sigma\},\{B_{\sigma\tau}\},Val_2)$ be R-structures. A system $\{f_\sigma\}$ is a <u>partial homomorphism</u> from M onto N just in case it is one when M,N are viewed as structures (the definition did not involve Val), and f_0 is the identity.

LEMMA 1. Suppose $\{f_\sigma\}$ is a partial homomorphism from M onto N. Suppose $x_1,\ldots,x_n$ are respectively in $Dom(f_{\tau_1}),\ldots,Dom(f_{\tau_n})$, and $A(x_1,\ldots,x_n) \in D^\sigma$. Then $f_\sigma(A(x_1,\ldots,x_n)) = B(f_{\tau_1}(x_1),\ldots,f_{\tau_n}(x_n))$.

<u>Proof</u>: By induction on n. For $n = 2$, this is straight from the definition of partial homomorphism. Using this, we have $f_\sigma(A(x_1,\ldots,x_{n+1})) =$ $= f_\sigma(A(A(x_1,x_2),\ldots,x_{n+1})) = B(f_\tau(A(x_1,x_2)),f_{\tau_3}(x_3),\ldots,f_{\tau_n}(x_n)) =$ $B(B(f_{\tau_1}(x_1),f_{\tau_2}(x_2)),f_{\tau_3}(x_3),\ldots,f_{\tau_{n+1}}(x_{n+1})) = B(f_{\tau_1}(x_1),\ldots,f_{\tau_n}(x_{n+1}))$ for appropriate τ.

The following Lemma is the analog to Lemma 16, for the R-calculus.

LEMMA 2. If $\{f_\sigma\}$ is a partial homomorphism from M onto N, and g is an M-assignment, h is an N-assignment, $f_\sigma(g(x)) = h(x)$ for variables x of type σ, then $f_\sigma(Val_1(s,g)) = Val_2(s,h)$, for R-terms s of type σ.

<u>Proof</u>: By induction on s, where $M,N,\{f_\sigma\}$ are fixed. The variable, application, and λ-abstraction cases of the induction are as in the proof of Lemma 15.

We have $f_0(Val_1(0,g)) = f_0(0) = 0 = Val_2(0,h)$.

We must show that $f_{00}(Val_1(N,g)) = Val_2(N,h)$. It suffices to show that for all $y \in \omega$, $f_0(A(Val_1(N,g),y)) = B(Val_2(N,h),f_0(y))$. Since f_0 is the identity, we have $f_0(A(Val_1(N,g),y)) = A(Val_1(N,g),y) = y + 1 = B(Val_2(N,h),y) =$ $B(Val_2(N,h),f_0(y))$.

Finally we must show $f_{\tau\tau}(Val_1(R_\sigma,g)) = Val_2(R_\sigma,h)$, where $\tau = (\sigma,(0,\sigma))$. It suffices to show that $f_\tau(A(Val_1(R_\sigma,g),y)) = B(Val_2(R_\sigma,h), f_\tau(y))$, for all $y \in Dom(f_\tau)$. It suffices to show $f_{0\sigma}(A(Val_1(R_\sigma,g),y,z)) = B(Val_2(R_\sigma,h),f_\tau(y),$ $f_\sigma(z))$, for all $y \in Dom(f_\tau)$, $z \in Dom(f_\sigma)$. Again, it suffices to show

$f_\sigma(A(Val_1(R_\sigma,y),y,z,k)) = B(Val_2(R_\sigma,h),f_\tau(y),f_\sigma(z),k)$, for all $y \in Dom(f_\tau)$, $z \in Dom(f_\sigma)$, $k \in \omega$. We show that this is true by induction on k. Note that $f_\sigma(A(Val_1(R_\sigma,g),y,z,0)) = f_\sigma(z) = B(Val_2(R_\sigma,h),f_\tau(y),f_\sigma(z),0)$. Assume true for k, and write $f_\sigma(A(Val_1(R_\sigma,g),y,z,k+1)) = f_\sigma(A(y,A(Val_1(R_\sigma,g),y,z,k),k)) =$ $= B(f_\tau(y),f_\sigma(A(Val_1(R_\sigma,g),y,z,k)),k) = B(f_\tau(y),B(Val_2(R_\sigma,h),f_\tau(y),f_\sigma(z),k),k) =$ $= B(Val_2(R_\sigma,h),f_\tau(y),f_\sigma(z),k+1)$, by Lemma 1, since $y \in Dom(f_\tau)$, $A(Val_1(R_\sigma,g),y,z,k) \in Dom(f_\sigma)$, $k \in Dom(f_0)$.

LEMMA 3. Suppose there is a partial homomorphism from M onto N, where M,N are R-structures. Then $M \models s = t$ implies $N \models s = t$, for any R-terms s,t.

<u>Proof</u>: Analogous to Lemma 16, using the previous Lemma.

LEMMA 4. Let M be any R-structure. Then there is a partial homomorphism from T_ω onto M.

<u>Proof</u>: A special case of (the proof of) Lemma 18.

LEMMA 5. The relation $T_\omega \models s = t$ is Π^1_1.

<u>Proof</u>: We claim that $T_\omega \models s = t$ if and only if for all countable R-structures M, $M \models s = t$.

Suppose $T_\omega \models s = t$. Then by Lemma 4, all R-structures M have $M \models s = t$.

Suppose not $T_\omega \models s = t$. Assume $T_\omega \models s \neq t[f]$. Then let M be a countable elementary substructure of T_ω containing $Rng(f)$, in the appropriate sense. M will be a countable R-structure, and $M \models s \neq t[f]$. So not $M \models s = t$.

We now wish to complete the proof that the relation $T_\omega \models s = t$ is complete Π^1_1.[2/] To this end, let P be the set of indices of primitive recursive well

---

[2/] This half of the proof was motivated by a proof by R. Gandy and G. Kreisel (Communicated to us by H. Barendregt) which showed that there are two unequal p.r. functionals which agree on all primitive recursive functional arguments.

orderings whose field is ω, whose least element is 0, and whose greatest element is 1. We can arrange the indexing so that every e is the index of a primitive recursive linear ordering $<_e$, whose field is ω, whose least element is 0, and whose greatest element is 1, and so that P is complete Π^1_1.

Let $(a_0,\ldots,a_n,\overline{0})$ be the function f given by $f(i) = a_i$ for $i \le n$; 0 otherwise.

Let $F\colon \omega^\omega \to \omega$. We wish to define $f = \Phi_e(F) \in \omega^\omega$. Let $f(0) = 1$. Let $f(n+1) = F((f(0),\ldots,f(n),\overline{0}))$ if $F((f(0),\ldots,f(n),\overline{0})) <_e f(n)$; 0 otherwise. Let $\Phi^*_e(F) = g$ be given by $g(n) = f(n)$ if $f(n+1) \neq 0$; 0 otherwise.

LEMMA 6. If $e \in P$ then for all F, $F(\Phi^*_e(F)) = F(\overline{0})$ or $F(\Phi^*_e(F)) \neq 0$.

Proof: Since $e \in P$, let $\Phi_e(F)$ be $(a_0,\ldots,a_n,\overline{0})$, where $n \ge 0$, $a_0 = 1$, and $a_0 >_e \cdots >_e a_n$, $a_n \neq 0$. If $n = 0$ then $\Phi^*_e(F) = \overline{0}$, and we are done. Otherwise, $\Phi^*_e(F) = (a_0,\ldots,a_{n-1},\overline{0})$. Hence $a_n = F(\Phi^*_e(F))$. So $F(\Phi^*_e(F)) \neq 0$.

LEMMA 7. For all e, $e \in P$ if and only if for all F, $F(\Phi^*_e(F)) = F(\overline{0})$ or $F(\Phi^*_e(F)) \neq 0$.

Proof: Assume $e \notin P$. Let $(a_0,a_1,a_2,\ldots)$ be such that $a_{i+1} <_e a_i$, and $a_0 = 1$. Clearly each $a_i \neq 0$, and so we may choose $F\colon \omega^\omega \to \omega$ such that $F(\overline{0}) = 1$, $F((a_0,a_1,a_2,\ldots)) = 0$, and $F((a_0,\ldots,a_n,\overline{0})) = a_{n+1}$ for $0 \le n$. Clearly, $\Phi_e(F) = \Phi^*_e(F) = (a_0,a_1,a_2,\ldots)$. Hence $F(\Phi^*_e(F)) = 0 \neq F(\overline{0})$.

For each e, we let ψ_e be the functional of type $(((0,0),0),0)$ given by $\psi_e(F) = 0$ if $F(\Phi^*_e(F)) = F(\overline{0})$ or $F(\Phi^*_e(F)) \neq 0$; 1 otherwise.

LEMMA 8. For all e, $e \in P$ if and only if ψ_e is constantly 0.

Proof: Obvious from Lemma 7.

LEMMA 9. There is a total recursive function α such that for each e, $\alpha(e)$ is a closed R-term of type $(((0,0),0),0)$ such that in T_ω, $\mathrm{Val}(\alpha(e)) = \psi_e$.

Proof: This just says that ψ_e is a primitive recursive functional, defined effectively from e .

THEOREM 6. The relation $T_\omega \models s = t$, for R-terms s,t , is complete Π_1^1 .

Proof: By Lemma 5, the relation is Π_1^1 . By Lemmas 8, 9, together with the fact that P is complete Π_1^1 , we see that the relation is complete Π_1^1 .

State University

of New York at Buffalo

REFERENCES

[1] H. Barendregt, Some extensional term models for combinatory logics and λ-calculi, Dissertation, University of Amsterdam.

[2] L. E. Sanchis, Functionals defined by recursion, Notre Dame J. Formal Logic, vol. 8, no. 3, pp. 161-174.

[3] W. Tait, Intensional interpretations of functionals of finite type, JSL 32 (1967), 198-212.

THE USE OF ABSTRACT LANGUAGE IN ELEMENTARY METAMATHEMATICS: SOME PEDAGOGIC EXAMPLES

G. Kreisel, G. E. Mints and S. G. Simpson

Introduction. 1. The logical need for abstract language is well-known, where by 'abstract language' we mean such things as the language of set theory or the many-sorted languages of higher type, in contrast to the 'concrete' languages of arithmetic or of concatenation theory . This logical need consists in the fact that there are (true) number theoretic and metamathematical propositions which can be proved by use of evident (and familiar) properties of abstract notions, but not by use of evident number theoretic or syntactic properties. Naturally, one needs an abstract language in the sense above to express properties of abstract notions.

Besides this logical need, there is also a practical or mathematical need for abstract languages, which has long been taken for granted; so much so that, in certain areas, for example, in current analytic number theory (where the abstract languages can be, demonstrably, eliminated 'in principle'), mathematicians tended to confuse logical and mathematical needs; specifically, many thought that analytic notions- of the complex plane and its structure - were needed logically, for the existence of some proof of the theorems considered, when in fact they were needed for intelligible proofs, of manageable complexity. Perhaps the most striking example of the distinction between logical and mathematical need is provided by the use of the axiomatic method applied, in, say, number theory. Here certain relations are explicitly defined and shown to satisfy more or less familiar 'axioms' in the theory of finite fields or topology.

Then results from these axiomatic theories are applied to the explicitly defined relations to yield number theoretic theorems . Since formal number theory is incomplete, there is of course the possibility that higher set theory is needed logically somewhere to verify that the explicitly defined relations satisfy the axioms in question or that the results are in fact consequences of the axioms. (This last verification is elementary by completeness if all formulae are of first order and if logical consequence is meant.) But these are only possibilities; in fact, they are not realized in current practice, even when there is general agreement on some kind of need for abstract methods. The obvious conclusion is that the need is simply not logical; no strong (existential) axioms for the abstract concepts are needed. For a very convincing analysis of the other kind of need, in terms of the measures of complexity provided by length and genus of derivations, the reader should consult Statman's dissertation [St].

Over the last two decades it has become clear that even elementary metamathematics benefits from the use of abstract language. As our title indicates, we wish to illustrate this view by use of pedagogic examples. But at least the thoughtful reader will demand a little more discussion of this switch since it conflicts, *prima facie*, with the principal aim for which Hilbert introduced metamathematics, that is, the aim of eliminating abstract language altogether. So it would at least appear circular to use such language in metamathematics itself. What has changed since the turn of the century when Hilbert formulated his program? For one thing some of us have come to doubt the particular hypotheses (about the nature of mathematical reasoning and hence of mathematical rigour) which suggested Hilbert's program - and, in fact, suggest that it would be easy to carry out. Those hypotheses do not appear even remotely plausible on the basis of intellectual experience; only on the basis of ideological, alias philosophical, views of what 'ought' to be reliable knowledge. But apart from this so to speak negative development, this correction of false first impressions, we also have a positive development in the present century: the discovery of natural and manageable 'models' of abstract languages, models to which we refer constantly in this paper. Here the reader should not forget that even the notion of hereditarily finite set has become really familiar only in the second half of this century! Otherwise it is hardly likely that Ackerman would have gone into as much detail as he did in [A] to discuss the axioms of general set theory excluding the axiom of infinity, or that Gödel would have arithmetized syntax instead of coding it in terms of hereditarily finite sets (where the representation of finite sequences of sets is quite straight-forward). – We invite the reader to recognize the defects of the past. There is quite enough of (logical) value in the ideas of the pioneers to make pious reverence quite inappropriate. Besides, any overestimate of achievements of the past necessarily implies an underestimate of progress since then.

Returning now to our immediate purpose, we naturally choose the most familiar metamathematical results for our pedagogic examples, namely completeness and so-called cut elimination theorems. The main results and especially the rôle of abstract language, in formulations and proofs, are described below in general terms. More details are to be found in Sections I and II.

2. Completeness of a system R of rules w.r.t. a formula A of a language L involves, obviously, the abstract notion of realization or model. To be quite precise, a realization is given by a domain (of individuals or several domains in the case of many sorted languages) and its satisfaction relation defined on some subclass of formulae in L. As is well-known in the particular case of first order predicate logic (here taken to contain no other function symbols than, possibly, constants for individuals), the satisfaction relation on atomic formulae determines the satisfaction relation for all formulae, and therefore the passage from the former to the latter is not analyzed. However, if – as mentioned in para. 1 – one is interested in 'restricted' classes of sets, e.g. recursive ones, the distinction becomes crucial: for example, the usual model for first order arithmetic in the language of rings has a recursive satisfaction relation when restricted to atomic formulae, but not even an arithmetically definable one for the whole language. As is to be expected from

experience in logic over the last decade, we shall consider different kinds of models, not only the usual binary ones; to avoid conflict with ordinary (mathematical) usage we shall then speak of valuations rather than 'models'. Actually, we shall not have occasion to use the familiar boolean valuations, but semi-valuations and partial valuations introduced by Schütte [Sch] some 15 years ago in connection with Takeuti's conjecture for inpredicative higher order logic. The partial valuations reformulated as ternary valuations were used quite recently by Girard to say so to speak the last word on this conjecture [Gi]. We shall use these valuations for our extensions of the completeness theorems for the usual rules with and without cut.

Summarizing: the notion A is valid or

A is true in all valuations, in symbols $\mathrm{Val}_\rho(\mathcal{L}, L', A)$

depends on the class $\mathcal{L}$ of sets considered, the kind ρ of valuations, and the subclass L' of formulae on which the valuations are given.

Now we use abstract language to reformulate the 'proof theoretic' side of completeness, namely

A is derivable by means of the rules R.

Instead of restricting ourselves to finite iterations of R, we consider formally (or 'pseudo') well founded trees (of formulae) where each node N is at a (literally) finite level and the formula at N is 'derived' by means of R from the formulae at the nodes immediately preceding N in the tree ordering; we shall also say that the trees are locally correct or locally regulated (by R). If N is terminal, the formula there must be an 'axiom'.

The formulation of 'A is derivable' in terms of well-founded trees is very familiar from ω-logic (where rules are in any case iterated transfinitely often); cf., for example, [KK] p.151. In the case of finitary rules, and thus of finitely branching derivation trees, formally well founded trees are finite provided the class $\mathcal{L}$ of sets or paths ensured by our formal theory is 'sufficiently rich'. It is not if both the trees and the paths considered are recursive since there are infinite recursive binary trees containing no infinite recursive path. Evidently, if $\mathcal{L}$ satisfies König's lemma* (and the trees considered are in $\mathcal{L}$) then our trees will be finite.

The extensions of the usual completeness theorems which we prove, all have the following form. For appropriate ρ (depending on R) and L' (depending on R and A)

$$A \text{ has a } \mathcal{L}\text{-founded R-tree if and only if } \mathrm{Val}_\rho(\mathcal{L}, L', A),$$

provided only $\mathcal{L}$ is closed under some simple primitive recursive operations. Actually, our results are a little sharper in the direction $\leftarrow$: there is a specific tree T_A, determined primitive recursively from the formula A, with the property:

$$\mathrm{Val}_\rho(\mathcal{L}, L', A) \rightarrow (T_A \text{ is locally regular and } \mathcal{L}\text{-founded}).$$

T_A is sometimes called the universal 'search tree' for A though, for our exposition, it is more appropriate to call it the universal refutation tree since it codes all infinite countable counter models to A (with domains consisting of terms familiar from Henkin's complete and consistent extensions).

Naturally, by what was said in the last paragraph but one, the extension is interesting or at least novel mainly if $\mathcal{L}$ does not satisfy König's lemma. For, if $\mathcal{L}$ does satisfy the lemma, the set

$$\{A : A \text{ has a } \mathcal{L}\text{-founded R-tree}\}$$

is obviously independent of $\mathcal{L}$; but so is

$$\mathrm{Val}_\rho(\mathcal{L}, L', A)$$

by a familiar use of the first 'basis' theorem in the literature (which concerns strict Π^1_1-predicates; this does not conflict with the fact, noted above, that a given satisfaction relation on L' in such a $\mathcal{L}$ cannot generally be extended to a satisfaction relation in $\mathcal{L}$ defined on the whole of L). So our reformulation follows from known, albeit useful facts about predicate logic. As to the interest of the extensions when $\mathcal{L}$ does not satisfy König's lemma (for example, if $\mathcal{L}$

---

*There are some technical distinctions concerning König's lemma which it is useful to keep in mind here; they are summarized in the Appendix.

consists of the recursive sets), there is nothing problematic if we are interested in recursion theoretic complexity of models; in this case our extension has the same kind of interest as does the usual completeness theorem for unrestricted models. There is an extensive literature on the subject culminating in papers of Vaught [V] or Jockusch-Soare [J-S]. Note that our results do not conflict with [V] since Vaught uses the satisfaction relation on atomic formulae as data (for 'giving' a model) which we never do for logically compound A.

However, it is possible to analyze the interest of our extension in a more sophisticated way relating it to general mathematical experience, expecially to the technique of studying distinct notions (here: rules) which are 'equivalent' in familiar domains D by extending the domain of definition to D^+ where they are not equivalent; or, in medieval terminology, where the notions are extensionally different. Here D refers to finite derivation trees, D^+ to $\mathcal{L}$-founded ones. We give such a sophisticated analysis at the end of Section I; not unexpectedly, it leads to novel problems very much in the spirit of (other) modern mathematics, for example geometry, if we compare rules to algebraic and $\mathcal{L}$-founded trees to analytic structures.

3. Rules with and without 'cut'. Inspection of the usual completeness proofs, at least when guided by the considerations of para. 2, provides the data, that is, the kind of valuation ρ and the class L' (depending possibly on A) where the satisfaction relation is given for which our extensions of the usual completeness theorems hold. Roughly speaking, for rules with cut, say R_1, ρ_1 is binary and L_1' consists of all formulae of the language (of A); for rules without cut, say R_2, ρ_2 is a semivaluation (defined precisely in I.2) and L_2' consists of a sufficiently large class of subformulae of A (and hence, in effect, on all formulae because a semivaluation on that class of formulae is a semivaluation on the whole language). Evidently

(*) $\{A : A$ has a $\mathcal{L}$-founded R_1-tree$\} = \{A : A$ has a $\mathcal{L}$-founded R_2-tree$\}$

if $\mathcal{L}$ satisfies König's lemma, simply because the cut free rules are complete in the usual sense and rules with cut are sound. (This provides the easy but perhaps not very well-known proof of 'cut elimination', for first order predicate logic, in a finite subsystem of first order arithmetic; cf. Appendix). However, both the general structure of so-called syntactic proofs of cut elimination and comparison with some crude considerations on ω-logic in Section II (p. 41) suggest that the rapid proof of (*) above misses the point! Specifically, we conjecture that (*) holds for some classes $\mathcal{L}$ for which *$\mathcal{L}$-founded trees are not automatically well-founded* (that is, $\mathcal{L}$ need not satisfy König's lemma, but only 'substantially weaker' closure conditions). – Naturally here we cannot expect the relevant conditions to be so *strikingly* weaker than (true) well-foundedness as they are in the case of ω-logic and other infinitary systems of rules: the latter are bound to magnify distinctions that are subtle in the case of ordinary first order predicate logic (a fact which, to one of us, is the main source of interest in infinitary languages).

Turning now to 'negative' results, that is, failure of (*), it is quite easy to see that (*) does not hold generally if $\mathcal{L}$ consists of the class of *recursive* sets. The easiest way to *find* a counter example is to look for a formula A which has no model with a recursive total valuation on the whole language of A but does have a recursive semivaluation (in which A is true). After having found such an A, it is usually quite easy to write down a Rec-founded derivation *with* cut of $\neg A$. Depending on one's practice with syntactic manipulations, one will or will not succeed in convincing oneself, by looking at possible cut free proofs, that $\neg A$ does *not* possess a Rec-founded derivation *without* cut. J. Stavi has improved on our original result by constructing a formula A which is true in some (primitive) recursive semivaluation, but in no recursive partial or, equivalently, ternary valuation (on the subformulae of A, and hence in no recursive binary valuation of the whole language). Precise definitions of these valuations are given in Stavi's exposition in I.2. His proof operates directly on the valuations without using the corresponding complete sets of rules (without cut, resp. with cuts limited to subformulae of A). We believe that Stavi's exposition has additional pedagogic interest as an *introduction to ternary valuations*, for example, as used by Girard [Gi].

Specifically, though Girard is careful to use 'weak' metamathematical methods, they are not weak enough to force a clear separation of notions, for example by analyzing the passage between partial (ternary) and semivaluations. (The Δ_2-formulae considered in his system AA^* are more than sufficient to define infinite paths of infinite recursive finitely branching trees).

NB. The notation in Stavi's subsection is different from the notation here; for example, he uses Greek l.c. letters for formulae while we use Roman capitals (for nearly everything).

4. Transformation of $\mathcal{L}$-founded trees; sharpening (*) in para. 3 by:(**) imposing (additional) requirements on mappings from R_1-derivations (of A) to R_2-derivations (of A), when (*) is true. In the literature the distinction between (*) and (**) is sometimes expressed by normal form and normalization theorems (when cut free rules are called 'normal'). For finite R_1-derivations the distinction was subtle because the property of being an R_1-derivation is decidable end so one mapping, ρ, is obtained trivially by simply running through all R_2-derivations till one hits an R_2-derivation with end formula A. From the start, there was little doubt about the interest of something like the particular normalization procedures in the literature though the analysis of that interest was problematic. It is fair to say that one of the most useful immediate consequences of the shift from (i) sequent formulations preferred by Gentzen himself and by Takeuti to (ii) natural deduction formulations preferred by Prawitz was just this: in contrast to (i), in case (ii) a particular mapping suggested itself; so there remained a well-defined normalization problem even after the corresponding normal form theorem had been established by model theoretic methods (which ensured at least one normalization procedure, the trivial ρ above). Five years later we cannot be satisfied with such virtues: quod decet bovem dedecet Jovem.

Once again we introduce abstract language, still the main theme of our

---

*On the other hand AA is 'too weak' for the particular applications made in [Gi]. For his purposes it is not necessary that the metamathematical proofs be formalized in AA; it would be sufficient to have them formalized in full classical analysis.

paper (though it has not been mentioned for some time) where the literature avoids it; specifically, we think of the derivation trees in (**) as arbitrary (sets), not as inductively defined objects, represented by natural numbers; in contrast, in the case of ω-logic, to Feferman [F] or Carstengerdes [C]. The most obvious additional requirement on ρ, suggested by this step is

continuity of ρ, where a neighborhood of a derivation is determined (as is usual for trees) by a finite number of finite initial paths in the given derivation tree.

Naturally, it remains to verify that this so to speak porte-manteau requirement, which makes sense throughout mathematics, is also of specific interest in the present context. This is plausible inasmuch as experience (of informal proofs) shows that we can often read off desired information from the last few steps of a proof without going all the way back to the beginning; for example, in classical mathematics if we want to know whether or not a given proof π of an existential theorem $\exists x B$ (with logically compound B) provides a specific (or even any) realization realization $x_\pi : B[x/x_\pi]$; cf. p.65 for more detail.

Another question which suggests itself immediately is this: can ρ be extended to a continuous total mapping, defined on all trees, not only R_1-derivations? There is a simple device for doing this, a kind of delay mechanism, familiar from recursion theory when one replaces Σ_1^1-well orderings by primitive recursive or even Kalmar elementary ones or partial recursive neighborhood functions by similarly elementary ones. This device is here applied by (i) adding to R_2, if necessary, the rule Repeat which allows a string of repetitions of the same formula on a (non-branching) path of an R_2-derivation and (ii) verifying that, if $\mathcal{L}$ is closed under recursive operations,

$$\{A : A \text{ has an } R_2\text{-derivation}\} = \{A : A \text{ has an } R_2^{Rep}\text{-derivation}\},$$

where 'R-derivation' means a tree which is $\mathcal{L}$-founded and regulated by the rules R.

Though the questions raised here are meaningful for $\mathcal{L}$-founded (finitely-branching) derivations in ordinary predicate logic, we discuss them, in II, for ω-logic: as was mentioned already, the distinctions are easier to see. In particular, at least as long as we consider ω-logic for finite formulae, the normal form theorem (*) requires now closure of $\mathcal{L}$ under all arithmetic operations, not the much weaker closure condition, of satisfying König's lemma, mentioned on p. 6. Conversely, it is certainly sufficient for (*) that $\mathcal{L}$ contains, with each X, the ω-th jump of X (in which all sets arithmetically definable from X are recursive): in this case any semi-ω-valuation, X, in $\mathcal{L}$ can be extended to a total (binary)-ω-valuation also in $\mathcal{L}$. Here it is evident that this closure condition is much weaker than the requirement that all

$\mathcal{L}$-founded ω-derivations are well-founded.

(This is the crude consideration on ω-logic mentioned in para. 3 above).

To conclude this discussion of transformations of derivations, let us note more generally that 'additional requirements' on normalization procedures ρ need a genuine study of proofs (not only of the set of provable theorems) if ρ is required to preserve some (structural) relations between proofs, for example, if, for such relations $\simeq$,

$$f \simeq_1 f' \Longrightarrow \rho f \simeq_2 \rho f \quad ^{*} .$$

As mentioned in para. 1 above, Statman's dissertation contains studies of such relations, defined in terms of simple isomorphism types of derivations.

Such requirements are to be distinguished from another type of additional requirement or 'sharpening' of (*), which can be perfectly adequately expressed in terms of sets of derivable theorems without explicit reference to derivations and their transformations. For example, our ρ (called ρ_1 in Section II) has the additional property that – except possibly for the rule Repeat – every rule applied in the derivation ρf is also applied (somewhere) on f ; no (applications of) rules are 'added', though some may be eliminated. However, in contrast to requirements of preserving relations between f and f', this second sharpening can be expressed quite adequately as follows. Let f_1 and f_2 range over trees of formulae and let $Der_1(f_1,A)$ stand for the conjunction

*One of us used precisely this phrase 'preserving relations between proofs' [SPT II, p. 100] but 5 pages later got involved in a discussion of the relation between (the proofs expressed by – what corresponds to –)f and ρf if ρ is the normalization operation for systems of natural deduction.

$$LC_i(f_i) \wedge WF(f_i) \wedge \lambda f_i = A$$

where LC_i means local correctness for the rules R_i, WF means well-foundedness, and λf_i is the end formula of f_i (which is of course determined by f_i, and so we need not quantify over A below). So the present 'sharpening' of (*), in particular, of

$$\forall f_1 \{Der_1(f_1, A) \rightarrow \exists f_2 Der_2(f_2 A)\}$$

is given by: $\forall f_1 \{Der_1(f_1, A) \rightarrow \exists f_2 [Der_2(f_2, A) \wedge R(f_1, f_2)]\}$ where R expresses that (except possibly for <u>Repeat</u>) all rules used on f_2 are also used on f_1. Clearly this is equivalent to the conjunction, over all suitably matching subsets R_i' of R_i $(i = 1,2)$, of

$$\{A: A \text{ has an } R_1'\text{-derivation}\} \subseteq \{A: A \text{ has an } R_2'\text{-derivation}\} \quad .$$

This involves only sets of <u>derivable</u> theorems, as stated earlier, not the derivations themselves.

<u>Remark</u>. One of us has advocated, starting in [SPT II] and continuing, perhaps <u>ad nauseam</u>, since then, that proof theorists give up the literally unrewarding aim of establishing essentially model theoretic results (about the set of deriv<u>able</u> theories) by proof theoretic manipulations (of derivations). Preservation results of the form: $f \simeq_1 f' \rightarrow \rho f \simeq_2 \rho f$ (where $f \simeq f'$ is <u>not</u> determined by λf and $\lambda f'$!) are good examples of genuinely proof theoretic results while the second kind of 'sharpening' is not. The conviction behind this propoganda over the last 5 years was that though traditional proof theory <u>looks</u> dull it was the statements of the results, not the methods used that were dull. In other words, the interest (of the methods) was real enough, but its analysis (expressed by the choice of theorems actually formulated) was defective. — Evidently those who pursue the subject for what were called ideological reasons cannot agree: no amount of structural analysis can compare, in glamor, with replacing dubious rules R_1 by valid ones, R_2!

Perhaps it should be pointed out that there are also quite different attempts (implicit) in the literature, of making existing proof theory more interesting; for example, Takahashi's imaginative idea [Ta2] that proof theory is concerned with the topological properties of the space of truth values. (This is certainly consistent with what most people know of the proof theory of classical logic with its discrete 2 element space of truth values). Somewhat perversely he chose continuous model theory [CK] to illustrate the general idea.

5. The need for enrichments of derivation trees: what is a proof tree? In view of the stress on continuity properties, of the mapping ρ in para. 4, we must be prepared to pay attention to the exact choice of data (which constitute the arguments of ρ, that is, the derivation trees on which ρ operates). It may fairly be said that a principal difference between the styles of modern mathematics and of a hundred years ago is just this: we have learnt to become aware of the effects of small differences in the choice of data and, consequently, can use those effects to lead us to a correct choice of data. In contrast, earlier workers attempted to evade the problem of choosing among 'slightly' different data by establishing brutal equivalences. – According to taste the reader may think of examples from category theory and the reason for enriching (the graphs of) functions by adding a bound on their range as part of the data; or from the theory of continuous functions on the unit interval and the reason for enriching (or 'equipping') such functions with a modulus of uniform continuity (to determine upper and lower bounds continuously on the uniform convergence topology); or from the theory of r.e. sets and the reason for equipping any such set with some enumeration to satisfy the reduction principle by mappings μ_1, μ_2 which are continuous for the product topology (on the enumerations, say e_x, e_y, of the sets X and Y): $\mu_1(e_x,e_y) \subset X \wedge \mu_2(e_x,e_y) \subset Y \wedge \mu_1 \cap \mu_2 = \emptyset \wedge \mu_1 \cup \mu_2 = X_1 \cup X_2$.

The principal criteria of choice of data will be to ensure that ρ can be chosen to be recursive and continuous and to have the further structural properties analyzed in para. 4. It will turn out that these criteria are only peripherally corrected with the main traditional metamathematical requirements which demanded data permitting quantifier-free proofs of (the usual) metamathematical results about transformations.

The data fall into two parts, (i) the presentation of the 'abstract' or 'underlying' trees and (ii) the information, besides the formulae 'proved', which is put at each node. As to (i), the basic choice is <u>between</u> a presentation of the set of all initial paths (without, for example, determining explicitly if a node is terminal, or more generally the exact number of its immediate predecessors) <u>and</u> one that includes the relation between nodes and explicit presentations of the set of immediate predecessors. The difference is very well illustrated in the Appendix in the case of binary trees which are given by a recursive set of initial paths (where each node 'happens' to have ≤ 2 predecessors) and those given by an additional recursive function defined on all nodes and taking as values a (code for) the set of predecessors, indicating here whether there are 2, 1, or none. In Section II we use the latter kind of presentation also in the case of infinitely branching trees. In short, at each node N we are given complete information about the <u>location</u> of the nodes used in the last inference which leads to (the formula at) N. —We think of this as <u>local</u> information about the 'underlying' tree because it refers only to nodes on the level adjacent to N.

As to (ii), the traditional literature on infinitary* proof figures introduces additional <u>global</u> information at each node N; partly about the underlying tree, namely a bound on the <u>ordinal length</u>, ℓ_N, of the tree T_N dominated by N, and partly about the formulae occurring in T_N, namely, a bound C_N on the logical complexity of either all formulae on T_N or at least of cut formulae (usually called <u>degree</u>). Though not exact values of length and complexity are required, the bounds chosen must be compatible with the tree ordering. – The immediate purpose for this additional information is of course clear,

---

*For finite proof figures this is rarely done because so to speak everything one wants to know <u>can</u> be read off from a finite figure. Realistically speaking, this is of course <u>not</u> true; we can often operate on a finite figure if we are given (relevant) additional information, but not if we have to search for it systematically.

especially of ℓ_N: modulo principles of transfinite induction formulated in the language of arithmetic (to which Gentzen's work had drawn the attention of proof theorists), this additional information allows us to paraphrase the basic condition of well-foundedness (WF on p. 11). A Π_1^0 - statement expresses the campatibility of the ordering of the notations ℓ_N with the tree ordering. The rôle of the bounds C_N was also clear for certain specific proofs of cut-elimination according to the lexicographic order (C_N, ℓ_N).

It should perhaps be mentioned that this additional information, though perfectly adequate for formulating the metamathematical arguments (especially of cut elimination) in the language of arithmetic, is not particularly convenient for a purely quantifier-free formulation: one is left with Π_1^0- proof predicates which appear as premisses in implications, for example, of the so-called reflection principle, even when applied to quantifier-free formulae. For partial results (possibly retaining quantifiers) a proper choice of extra information at each node is due to Lopez-Escobar [LE] who puts, at N, a suitable formal derivation, say d_N, of the formula 'proved' at N (where d_N is in the system in which the metamathematical result in question is to be formalized); more detail on this kind of enrichment will be found in Section II.2(c).

Even if, subjectively, the additional information ℓ_N, c_N, d_N was ad hoc, that is, was actually obtained by tinkering, it is pretty clear that the information is pertinent to the aim of 'arithmetizing' metamathematics in an elementary way. We do not question this here. What we wish to study is this: Is this information on which we have stumbled in pursuing the quite different aim (of giving quantifier-free proofs of metamathematical results), also useful for our present more structural aims? More precisely,

> Must, for example, ω-derivations f be enriched by bounds (c_N, ℓ_N) to solve the 'functional equation' (for ρ):
>
> $\mathrm{Der}_1(f_1,A) \to \mathrm{Der}_2(\rho f_1,A)$, by means of a continuous ρ?

The answer, given in Section II, is negative. Next we ask:

> Must we make a fresh start to solve the functional equation

above when the f_1 are enriched by (c_N, ℓ_N)? (And correspondingly, ρf_1 is to be enriched too.)

Again, the answer is negative, since thereis a mapping β defined on the 'underlying' (universal) trees which maps bounds into bounds, compatible with the tree ordering, and β is independent of the formulae on f_1. As one might expect, β is closed under ϵ_0, that is, if $\ell_N < \epsilon_0$ for all N so is (the second component of) $\beta(c_N, \ell_N)$. In other words, the business of bounds is not of the 'essence', but can be separated from the structural analysis.

6. Conclusion. Evidently our analysis of the data (determining derivations) and of requirements on operations, ρ, is quite superficial compared to the corresponding work in venerable parts of mathematics, for example, geometry. There the data, determining geometric figures (in terms of the 'structure' of the space: combinatorial, algebraic, differential, etc) are analyzed in an incomparably more sophisticated way, and the same applies to various kinds of 'equivalences' and 'isomorphism types'. But it seems to us that our analysis has at least the right smell.

I. GENERALIZATIONS OF THE COMPLETENESS THEOREM FOR FIRST ORDER PREDICATE LOGIC.

In subsection 1 we sketch what is, perhaps, the most basic case where (i) the models considered have as domains terms familiar from Henkin's construction of complete and consistent extensions, and a model (of A) is determined by the satisfaction relation for all formulae in the language of A, that is, built up from the relation symbols occurring in A, and (ii) the rules of inference include cut. The reader should have no difficulty in working out the next most familiar case when (i') the models considered are determined by semi valuations in the sense described precisely in Section 2, and (ii') the rules of inference exclude cut.

In subsection 2, most of which was written by J. Stavi, a formula A is constructed which is (i) true in every recursive partial or ternary valuation, and a fortiori in every recursive complete and consistent extension (of the language of A) but (ii) false in some (primitive) recursive semi-valuation. Thus, by subsection 1, A has a Rec-founded derivation with cut, but none without cut.

In subsection 3, there is a discussion of possible uses of our extensions or 'generalizations' of familiar completeness theorems, with special stress on new problems formulated in terms of (the notions used in) the generalization. We can hardly hope that these new problems, the product of research over the last 40 years, can have the same immediate appeal as the original completeness problem for logical validity which, after all, is intelligible without any specialized knowledge! So both for objective and for subjective pedagogic reasons we go into the limitations of the original problem; so to speak its insensitivity to the class of models considered (which was a great advantage for preliminary orientation). Naturally, as is familiar from other parts of science, what were mere auxiliaries in the solution of the original problem turn out to be rewarding principal topics of research, and by no means of less permanent interest than the original problem. – As examples of interesting (and precise) problems we should mention here two kinds of 'basis' problems: (i) for a given kind (ρ, L') of valuation (cf. p. 4) to find $\mathcal{L}_{(\rho,L')}$ s.t. for all A: $\mathcal{L} \supseteq \mathcal{L}_{(\rho,L')} \Rightarrow [\mathrm{Val}_\rho(\mathcal{L}, L', A) \Leftrightarrow \mathrm{Val}_\rho(\mathcal{L}_{(\rho,L')}, L', A)]$ and (ii) for a given pair $(\rho_1, L_1'), (\rho_2, L_2')$ to find $\mathcal{C}_{(1,2)}$ such that, for $\mathcal{L}$ satisfying conditions $\mathcal{C}_{(1,2)}$, $\forall A[\mathrm{Val}_{\rho_1}(\mathcal{L}, L_1', A) \Leftrightarrow \mathrm{Val}_{\rho_2}(\mathcal{L}, L_2', A)]$.

1. Universal Refutation Trees: Model Theoretic Analysis and the Search for Complete Sets of Rules.

Historically, (Frege's) rules came first and their completeness (for validity) was proved some fifty years later (by Gödel). One so to speak started with the 'linguistic evidence', and slowly passed to the (model theoretic) 'semantic' notions consistent with this 'evidence'. We reverse this procedure here, starting with the semantic notions of model and validity. To some of us, those notions, once isolated, are clearer and, above all, more obviously significant than any particular set of rules can ever hope to be.

We begin with quite crude distinctions, for example, in terms of the <u>cardinality</u> of models and only afterwards analyze more precisely the closure conditions on the class of models actually used.

(a) Cardinality: validity for arbitrary models is equivalent to validity for countable models. (If A has an uncountable model M there is a countable elementary submodel of M in which A is true. If A has a finite model M_0 with domain $a_1,\ldots,a_N$, A has a countable model M_0^+ whose domain is the union of copies $M_0^{(n)}$ of M_0 and each $a_i^{(n)}$ is regarded as a distinct <u>name</u> of a_i:

$$R(a_{i_1}^{(n_1)},\ldots,a_{i_k}^{(n_k)}) \text{ is true in } M_0^+ \text{ iff } R(a_{i_1},\ldots,a_{i_k}) \text{ is true in } M_0.$$

The passage from M_0 to M_0^+ uses the fact that we consider general models and not 'normal' or 'equality' models.)

(b) Term models, also called canonical models. Validity for countable models is also equivalent to validity in all countable term models of the particular kind introduced by Henkin (for exposition, see [KK], p. 83, where the language L_Δ is described). Here, if A is a formula with the single variable x and $\exists xA$ is true in the model so is $A[x/\epsilon_A]$ where ϵ_A are the only constants. If A is false, e.g. of the form $B \wedge \neg B$, then no requirement is imposed on ϵ_A. The canonical models considered have as domain the terms ϵ_A where $A \in L_\Delta$. The reader will find relations to other kinds of canonical models in Ex. 2 on p. 105 of [KK]).

The role of these particular term models for a study of the <u>recursion theoretic complexity</u> of models is clear. Suppose our data, determining a model, consist of its domain and its satisfaction relation (or, equivalently, the truth definition for a language augmented by names for all elements in the domain of the model), and suppose the closed formula $\exists xA$ is true, that is, <u>belongs</u> to the satisfaction relation. Then, in general, we need a <u>search</u> through $A[x/c]$ where c are names of the elements of the model to find one, say c_A, such that $A[x/c_A]$ is true. For our canonical models this operation is more <u>primitive</u>.

(c) At this stage we set ourselves the problem of <u>coding all countable canonical counter models to a given formula</u> A, or, equivalently, such models of $\neg A$. (The decision to look at <u>counter</u> models, in the case of classical logic, is based on the familiar fact that validity here has nothing to do with explicit constructions: one constructs counter examples to proposed refutations). The strategy for this coding procedure is by now familiar, for example, from so-called <u>semantic tableaux</u> (although, in contrast to our aim, these tableaux provide only a certain class of semivaluations, on <u>subformulae</u> of A, in which A is false). We describe a binary tree T_A, depending primitive recursively* on A, at whose nodes N we have two sets of formulae (Γ_N, Δ_N) or, more often, $\Gamma_N \vdash \Delta_N$, with the following properties.

If $\langle\rangle$ is the root of T_A, $\Gamma_{\langle\rangle} = \emptyset$, $\Delta_{\langle\rangle} = \{A\}$.

Each infinite path π through T_A <u>determines</u> a <u>countable canonical model</u> $\mathfrak{M}_\pi$ of $\neg A$ where the true formulae of $\mathfrak{M}_\pi$ are given by $\{F: F \in \Gamma_N, N \in \pi\}$ and the false formulae by $\{F: F \in \Delta_N, N \in \pi\}$; thus, for $N, N' \in \pi$, $\Gamma_N \cap \Delta_{N'} = \emptyset$ and $\bigcup \{\Gamma_N \cup \Delta_N : N \in \pi\}$ must be the whole language.

Each countable canonical model M of $\neg A$ <u>determines</u> one or more infinite paths π_M through T_A for all of which

$$\mathfrak{M}_{\pi_M} = M .$$

The <u>extension</u> or 'generalization' stressed in this paper comes from analyzing a little the operations

(*) $\mathfrak{M}: \pi \longmapsto \mathfrak{M}_\pi$ and: $M \longmapsto \pi_M$;

to make them as elementary or explicit as possible, one must pay attention to the specific construction of T_A. As a result we have this.

<u>Corollary</u>: Let $\mathcal{L}$ be any class of sets closed** under the operations (*);

---

* Or indeed, in a (Kalmar-) elementary way.

** As far as 'Gödel numberings' are concerned we naturally require them to satisfy the familiar conditions on canonical numberings (which are isomorphic by means of elementary operations). Our $\mathcal{L}$ are to be closed under those too.

the paths which are in $\mathcal{L}$ and go through T_A correspond to models of $\neg A$ given by data in $\mathcal{L}$. – Another

Corollary: A is true in all models in $\mathcal{L}$ if and only if T_A is $\mathcal{L}$-founded.

Remark. The reader who can see in outline the (familiar) construction of T_A should go on immediately to (d) on p.20 where we pass from this construction to the discovery of rules of inference.

Construction of T_A (for A belonging to L, that is, not containing the constants ϵ_F for $F \in L_\Delta$). We enumerate all closed formulae A_i of L_Δ in an order respecting subordination of terms. At stage $2i$ $(i > 0)$ each possibility (A_i true, resp. A_i false) will be 'envisaged' in a branching of T_A to make sure that each path contains A_i either to the left or to the right of $\vdash$ at some node. At stage $(2i+1)$, which is broken up into a finite number of 'logical' analyses, the internal (logical) structure of the formulae in Γ_N, Δ_N (N at stage $2i$) is 'considered'. Specifically, suppose our language is

$$\neg, \vee, \exists;$$

and let N_j $(j \leq \rho(i))$ be the nodes introduced at stage $2i$. (The reader who likes number theoretic functions may compute ρ; paying attention to the fact that at each odd stage several formulae are analyzed and so the depth of T_A say, T_A^{2i+1} increases by > 1 between T_A^{2i} and T_A^{2i+1}.) The propositional analysis is standard:

If at the node N_j we have $\{\neg F\} \cup \Gamma' \vdash \Delta$ or $\Gamma \vdash \{\neg G\} \cup \Delta'$, and $\neg F$, resp. $\neg G$ is being analyzed then N_j has just one immediate predecessor: $\Gamma' \vdash \Delta \cup \{F\}$, resp. $\Gamma \cup \{G\} \vdash \Delta'$. If we have $\Gamma \vdash \{F \vee G\} \cup \Delta'$, the immediate predecessor is $\Gamma \vdash \Delta' \cup \{F,G\}$. If we have $\{F \vee G\} \cup \Gamma' \vdash \Delta$ there is a branching: $\{F\} \cup \Gamma' \vdash \Delta$ and $\{G\} \cup \Gamma' \vdash \Delta$.

If at N_j we have $\{\exists xF\} \cup \Gamma' \vdash \Delta$ then (sometime during stage $2i+1$) $\exists xF$ is 'considered' and replaced by $F[x/\epsilon_F]$, in accordance with the requirement that if $\exists xF$ is true so is $F[x/\epsilon_F]$. If we have $\Gamma \vdash \{\exists xF\} \cup \Delta'$ then the formulae $F[x/\epsilon_k]$ for $k \leq i$ are added to the right of $\vdash$; put differently at the terminal nodes N of T_A^{2i+1} dominated by N_j, $\Delta_N \supset \{F[x/c_k]:k \leq i\}$ for the first i constants c_k of the language L_Δ. This makes sure that if $\exists xF \in \Delta_N$ on some node N of an infinite path π through T_A then each $F[x/\epsilon_G] \in \bigcup \{\Delta_N : N \in \pi\}$, where $G \in L_\Delta$.

STOP the construction at node N if $\Gamma_N \cap \Delta_N \neq \emptyset$.

What is there to prove? First of all, each path π clearly determines a model (that is, complete and consistent extension) of $\neg A$, say $\mathfrak{M}_\pi$: the truth value of the ith formula A_i being read off from the position of A_i at the node of π introduced at stage $2i$ (completeness). For consistency, argue by induction on logical complexity. The mapping $\mathfrak{M}$ is quite explicit because we know where to look for A_i on π. Secondly, a counter model M to A determines one or more infinite paths through T_A by telling us which branching to choose at each node: if N_1 and N_2 are the immediate predecessors of N, and Γ_N are all true in M, Δ_N are false then Γ_{N_i} must all be true, Δ_{N_i} must all be false (in M), for $i=1$ or $i=2$ or both.

At stage $2i$ ('cut') there is no choice: if A_i is true in M, we must take that branch which has A_i on the left of $\vdash$, and conversely. At stage $(2i+1)$, if $F \vee G \in \Gamma_{N_j}$ and so $F \vee G$ is true in M, the choice is open if $F \wedge G$ is true. Nevertheless

$$\mathfrak{M}_{\pi_M} = M$$

since the formulae of π at the nodes introduced at even stages determine the model.

(d) From refutation trees, T_A, to rules of inference (for A). At terminal nodes of T_A we have

$$\Gamma' \cup \{F\} \vdash \{F\} \cup \Delta' \quad \text{since } \Gamma \cap \Delta \neq \emptyset\text{, for ex., } F \in \Gamma \cap \Delta$$

where the order of occurrences of formulae in Γ or Δ is neglected.

At stages $2i$, the reversal of the construction of T_A is the rule of cut: from $\Gamma \cup \{A_i\} \vdash \Delta$ and $\Gamma \vdash \Delta \cup \{A_i\}$ infer $\Gamma \vdash \Delta$.

At stages $2i+1$, we have the usual rules for $\neg, \vee$; the 'usual' contraction rule for the existential quantifier ($\vdash\exists$)

from $\Gamma \vdash \{F[x/c], \exists xF\} \cup \Delta'$ infer $\Gamma \vdash \{\exists xF\} \cup \Delta'$

and the 'unusual' rule ($\exists\vdash$):

(*) from $\{F[x/\epsilon_F]\} \cup \Gamma' \vdash \Delta$ infer $\{\exists xF\} \cup \Gamma' \vdash \Delta$;

the 'usual' rule ($\exists\vdash$) corresponding to (*) requires, as premise, $\{F[x/c]\} \cup \Gamma' \vdash \Delta$ where c does not occur anywhere in $\Gamma' \cup \Delta$. To show then that the usual rule is sufficient to replace (*) we use the familiar substitution or

Standardization Lemma. Let L be the language of A expanded by the terms $c_0, c_1, \ldots$ (in place of the terms ϵ_F of L_Δ), and let T be a locally correct, but not necessarily well-founded derivation tree for A, regulated by the usual rules. Then there is a standard tree T' such that, for any two inferences, at nodes N, N', of

$$\{\exists xF_N\} \cup \Gamma'_N \vdash \Delta_N \text{ from } \{F_N[x/c_N]\} \cup \Gamma'_N \vdash \Delta_N$$

c_N and $c_{N'}$ are distinct. (In T' some 'eigen' variables of T may be renamed.)

Corollary. If T^1 is a derivation, by the usual rules, of a formula A not containing constants, then A has a derivation by the unusual rules too.

The corollary follows if, throughout a standard form of T^1, c_N is replaced by ϵ_{F_N}, since this replacement does not disturb the logical connections locally. As to the proof of standardization, the only point to notice is that T is not assumed to be well-founded, and so we do not use a construction by (transfinite) recursion, but proceed from the root of the tree T. We reserve the terms c_{2j} (with even subsript) for the jth node N at which $\{\exists xF_N\} \cup \Gamma'_N \vdash \Delta_N$ is inferred from $\{F_N[x/c_N]\} \cup \Gamma'_N \vdash \Delta_N$ on T' and replace c_N, which does not occur in $\Gamma'_N \cup \Delta_N$, by c_{2j} on all nodes N_1 'below' N if c_N is not bound between N_1 and N. Then terms c_i, at a node N' of T, which differ from all c_N occuring on the path joining N' to the root of T are replaced by c_{2i+1}. Since each node of T is, by definition, at a finite destance from the root, we do not run out of constants.

Exercise. The reader may wish to verify directly that the usual rules are sound for validity on the term models described in (b) above; that is, if A has a $\mathcal{L}$-founded tree regulated by the usual rules, A is true in every term model determined by data in $\mathcal{L}$-Hint. Some care is needed in the choice of data determining a tree; for more detail, see II.1(a).

(c) Concerning closure conditions on $\mathcal{L}$, it is of course evident that, if the trees T considered $\in \mathcal{L}$ and $\mathcal{L}$ satisfies König's lemma (in the usual sense, discussed in the Appendix) then T is finite, and hence the end formula of T is logically valid. The significance of these facts for the (unexpected) rôle of logical validity in mathematical reasoning is well-known; we shall return to this matter in subsection (3) below, b(i) on p.32.

2. $\mathcal{L}$-founded Derivations With and Without Cut: A Comparison.

As mentioned in the summary, at the beginning of the present Section I,

A is not false, that is, true or undefined, in each semi-valuation $\in \mathcal{L}$ if and only if a certain 'refutation' tree T_A^{CF}, regulated by rules without cut, is $\mathcal{L}$-founded. The definition of 'semi-valuation' is given below and also in [Sch] which contains a proof of completeness for semi-valuations (from which a proof for our generalization to $\mathcal{L}$-founded derivations can be extracted if one looks for it). Before passing on to the main results the reader may wish to review his knowledge of proof theory in terms of the generalization.

Exercise. Find weak closure conditions on $\mathcal{L}$ (for example, closure under primitive recursive operations) which are sufficient for so-called inversion theorems such as

(for conjunction) if $\Gamma \vdash A \wedge B, \Delta$ has a (cut free) $\mathcal{L}$-founded derivation then both $\Gamma \vdash A, \Delta$ and $\Gamma \vdash B, \Delta$ have such derivations; or

(existential premis) if the variable x does not occur in $\Gamma \cup \Delta$: $\exists x A, \Gamma \vdash \Delta$ has a cut-free $\mathcal{L}$-founded derivation iff $A, \Gamma \vdash \Delta$ has one.

The corresponding results are patently true for $\mathcal{L}$-founded derivations with cut; (for all $\mathcal{L}$ considered in subsection 1); model theoretically because all valuations are defined on all formulae of the language and the equivalences are true for validity; proof theoretically because, e.g. in the case of conjunction, we can attach to a derivation of $\Gamma \vdash \{A \wedge B\} \cup \Delta$ a (finite) derivation of $A \wedge B \vdash A$, resp. $A \wedge B \vdash B$ and <u>cut</u> with $A \wedge B$. – In the case of cut-free derivations a little care is needed; model theoretically, because we have to choose semi-valuations with suitably matching domains of definition; proof theoretically, because we cannot apply without restriction (on the predicates used) the principle of <u>induction</u> on the length of derivations since our derivations are only $\mathcal{L}$-founded. So we must either pay attention to the complexity of the predicate to which the principle is applied, or simply use a proof which starts with the end formula, not the terminal nodes of the derivation.

Returning now to our trees T_A of subsection 1 and T_A^{CF} we have the relation:

If $\mathcal{L}$ (is closed under, say, primitive recursive operations and) satisfies König's lemma

(**) $\qquad T_A^{CF}$ is $\mathcal{L}$-founded if and only if T_A is $\mathcal{L}$-founded,

or, equivalently, the same formulae have $\mathcal{L}$-founded derivations with and without cut.

The proof of (**) is trivial <u>modulo</u> the well-known cut-elimination theorems for <u>finite</u> derivations. (The reader may also give a new model theoretic proof showing that if A is true in some semi-valuation in $\mathcal{L}$ then A is also true in some* total binary $\mathcal{L}$ valuation of the whole language of A). For reasons considered in subsection 3 below, it is plausible that the equivalence (**) above holds under weaker closure conditions; in other words, (**) has 'nothing' to do with the equivalence between $\mathcal{L}$-foundedness (of binary trees) and well-foundedness.

It will now be shown that (**) is false (for suitable A) if

$\mathcal{L}$ is the class of recursive sets,

by showing that A is valid in all recursive total valuations but not in all recursive semi-valuations. Actually even more is true (for the A considered):

---

*but the given semi-valuation need not have an extension, to a total valuation, which is also in $\mathcal{L}$.

A is valid in all recursive valuations defined on all subformulae of A; thus the recursive semi-valuations in which A is not true cannot be defined on all subformulae of A.

The remainder of this section was written by J. Stavi; we have retained his notation which differs from ours.

An Example of Satisfiability by Recursive Semi-valuations

A. Basic Notions

Definitions: (a) We regard 0,1,2 as the three "truth values" – false, undetermined and true resp. For each (propositional) operation $*$, let $*_2$ be the ordinary truth function on $\{0,2\}$ associated with $*$ (we take the operations to be $\neg, \wedge, \vee, \rightarrow, \top, \bot$) and $*_3$ the natural extension of $*_2$ to $\{0,1,2\}$:

$$\top_3 = 2, \quad \bot_3 = 0,$$

$$\neg_3(x) = 2 - x,$$

$$\wedge_3(x,y) = \min(x,y)$$

$$\vee_3(x,y) = \max(x,y)$$

$$\rightarrow_3(x,y) = \vee_3(\neg_3(x),y) = \max(2-x,y) .$$

(b) Let L be a set of relation symbols, C a nonempty set of constants, Sent the set of sentences of the language L(C) (built up by means of the operations above and of the quantifiers $\forall$, $\exists$).

A total valuation, for L(C), is a function $V : \mathrm{Sent} \rightarrow \{0,2\}$ such that for every $\varphi \in \mathrm{Sent}$:

(i) If $\varphi = *(\varphi_1,\ldots,\varphi_n)$ (where $0 \leq n \leq 2$ and $*$ is an n-ary operation) then $V(\varphi) = *_2(V(\varphi_1),\ldots,V(\varphi_n))$.

(ii) If $\varphi = \forall x\psi(x)$ then $V(\varphi) = \min_{c\in C} V(\psi(c))$, and if $\varphi = \exists x\psi(x)$ then $V(\varphi) = \max_{c\in C} V(\psi(c))$.

A ternary valuation is a function $V : \mathrm{Sent} \rightarrow \{0,1,2\}$ such that, for every $\varphi \in \mathrm{Sent}$, (i) and (ii) hold, with $*_2$ replaced by $*_3$ in (i).

Remark: Total valuations and ternary valuations are uniquely determined by their restrictions to the atomic sentences, and every function on the atomic sentences into $\{0,2\}$ ($\{0,1,2\}$) can be extended to a total (ternary resp.) valuation. Thus, these valuations correspond biuniquely to (two-valued and three-valued) C-models for L(C), that is, a model with domain C such that every constant in C denotes itself. Of course, we are talking about models in which the equality symbol (if present in L at all) is not assigned a special realization.

Definitions (cont.): (c) A semi-valuation is a partial function $V:\text{Sent} \to \{0,2\}$ such that the following "downward" conditions hold; where D = domain of V, $D_i = \{\varphi\in D \mid V(\varphi) = i\}$ $(i = 0,2)$.

(1) For any $\varphi\in D_2$ one of the following is the case:

(i) $\varphi = \top$;

(ii) $\varphi = \neg\psi$ and $\psi\in D_0$;

(iii) $\varphi = \chi\wedge\psi$ and $\chi\in D_2$ and $\psi\in D_2$;

(iv) $\varphi = \chi\vee\psi$ and $\chi\in D_2$ or $\psi\in D_2$;

(v) $\varphi = \chi\to\psi$ and $\chi\in D_0$ or $\psi\in D_2$.

(2) For any $\varphi\in D_0$ one of the following is the case:

(i) $\varphi = \bot$;

(ii) $\varphi = \neg\psi$ and $\psi\in D_2$;

(iii) $\varphi = \chi\wedge\psi$ and $\chi\in D_0$ or $\psi\in D_0$;

(iv) $\varphi = \chi\vee\psi$ and $\chi\in D_0$ and $\psi\in D_0$;

(v) $\varphi = \chi\to\psi$ and $\chi\in D_2$ and $\psi\in D_0$.

(d) The set $\text{Sub}(\varphi)$ of subsentences of φ is defined recursively by:

$\text{Sub}(\varphi) = \{\varphi\}$ for atomic φ ;

$\text{Sub}(\varphi) = \{\varphi\} \cup \text{Sub}(\varphi_1) \cup \cdots \cup \text{Sub}(\varphi_n)$ for $\varphi = *(\varphi_1,\ldots,\varphi_n)$;

$\text{Sub}(\varphi) = \{\varphi\} \cup \bigcup_{c\in C} \text{Sub}(\psi(c))$ for $\varphi = \forall x\psi(x)$ or $\exists x\psi(x)$.

(e) A full semi-valuation for φ is a semi-valuation whose domain includes $\text{Sub}(\varphi)$.

(f) We abbreviate "semi-valuation" by "s-val.", "ternary valuation" by "3-val.".

Remark: The important features of the definition of a s-val. are: (1) The domain need not be closed under subsentences (if $\chi \vee \psi$ is true (ϵD_2) then one of χ, ψ must be true but the other may be undetermined $(\notin D)$).

(2) The requirements go only "downward". If χ and ψ are both true (ϵD_2) then of course $\chi \wedge \psi$ cannot be false but it need not belong to the domain D.

If we change the definition by adding the natural "upward" conditions ($\chi, \psi \epsilon D_2 \Longrightarrow \chi \wedge \psi \epsilon D_2$; $\chi \epsilon D_0 \Longrightarrow \chi \wedge \psi \epsilon D_0$ etc.) we obtain the notion of a partial valuation; the reader may check that the partial valuations in this sense are just the functions of the form $V | \{\varphi | V(\varphi) \neq 1\}$ where V ranges over 3-val.'s. Thus, partial valuations correspond biuniquely to 3-val.'s in a very simple way.

B. Statement of the Results

Suppose now that L is finite and its relation symbols are listed in a finite sequence, and suppose $C = \{c_n | n < \omega\}$ where $c_0, c_1, c_2, \ldots$ are distinct constants. If we choose a (primitive) recursive Gödel numbering of the constants, and distinct numbers for the other symbols, a Gödel numbering for all formulas of $L(C)$ is determined uniquely up to (primitive) recursive isomorphism by familiar conditions. Thus we can legitimately talk of (primitive) recursive valuation functions of all kinds.

Theorem: For a suitable choice of L the following hold: (a) There is a sentence α which is false under every recursive total valuation but is true under some prim. rec. full s-val. for α (see def. (e)).

(b) There is a sentence β which is true under some prim. rec. s-val. but not true under any recursive full s-val. for β and not true under any 3-val. (or partial valuation) whose restriction to $\mathrm{Sub}(\beta)$ is recursive.

Remark: We have also proved this for a language L which contains two binary predicates $(=, \epsilon)$; probably one binary predicate (ϵ) is enough. The proof below, however, will take for L the language of arithmetic (for ordered rings) except that $0, S, +, \cdot$ are replaced by the relation symbols zero, suc, sum, prod. For this language the proof follows directly from some well-known properties of "Robinson's arithmetic" or its variants.

C. Proof of Part (a) of the Theorem

Let L consist of the relation symbols =, zero, suc, sum, prod, < (with 2,1,2,3,3,2 arguments resp.). The sentence α will be the conjunction of the following sentences of L:

$\alpha_1 - \alpha_3$: Reflexivity, symmetry and transitivity of = .

$\alpha_4 - \alpha_8$: Substitutivity of = in the other predicates [e.g. α_6 is

$$\forall xyzx'y'z' (\{x = x' \wedge y = y' \wedge z = z' \wedge \text{sum}(x,y,z)\} \rightarrow \text{sum}(x',y',z')).]$$.

$\alpha_9 - \alpha_{12}$: single-valuedness axioms for zero, suc, sum, prod. [e.g. α_{10} is –

$$\forall xyz (\{\text{suc}(x,y) \wedge \text{suc}(x,z)\} \rightarrow y = z)]$$.

$\alpha_{13} - \alpha_{16}$: Existence of zero, successor, sum, product. [e.g. α_{15} is $\forall xy \exists z \smile \text{sum}(x,y,z)$].

$\alpha_{17} - \alpha_{25}$: The axioms $N_1 - N_9$ of the weak number theory N given in Shoenfield [Sh, p. 32], with 0,S,+,· eliminated in favour of zero, suc, sum, prod. Thus $\forall x(Sx \neq 0)$ becomes $\forall xy \neg(\text{zero}(y) \wedge \text{suc}(x,y))$ and $\forall xy\ (x\cdot Sy = x\cdot y+x)$ becomes – $\alpha_{13}-\alpha_{16}$ –

$$\forall xyzuvw (\{\text{suc}(y,z) \wedge \text{prod}(x,z,w) \wedge \text{prod}(x,y,u) \wedge \text{sum}(u,x,v)\} \rightarrow w = v).$$

The exact choice of translation is somewhat arbitrary. We use only, as in the two examples given, the fact that each of $\alpha_{17} - \alpha_{25}$ is a π_1 sentence (i.e. in the form $\forall v_1,\ldots,v_n\varphi$ where φ is a quantifier-free formula).

Thus $\alpha = (\cdots((\alpha_1 \wedge \alpha_2) \wedge \alpha_3) \wedge \cdots) \wedge \alpha_{25})$, or $\alpha_1 \wedge \cdots \wedge \alpha_{25}$ for short. Inspection shows that each α_i, except the existence axioms $\alpha_{13} - \alpha_{16}$, is a π_1 sentence. $\alpha_{13} - \alpha_{15}$ have the form $\forall v_1 \cdots v_k \exists u\varphi$ where k = 0,1 or 2 and φ is quantifier free.

To show that α has the desired properties we consider first the standard model

$\Omega = \langle\omega, =, \{0\}, \{(m,n)\,|\,n=m+1\}, \{(\ell,m,n)\,|\,n=\ell+m\}, \{(\ell,m,n)\,|\,n=\ell\cdot m\}, <\ ;\ 0,1,2,3\ldots\rangle$

for $L(C)$ (where c_n is n). Let V_Ω be the total valuation function associated with this model. Since $\Omega \models \alpha$, $V_\Omega(\alpha) = 2$.

To establish the positive part of theorem we show first that the restrictions of V_Ω to certain classes of simple sentences are prim. rec. Then we shall make use of well-known properties of N to prove the negative statement about α.

Lemma 1: The restriction of V_Ω to the set of all quantifier free sentences is prim. rec.

Lemma 2: Suppose the $L(C)$-sentence φ is true in Ω and has either of the forms $\forall v_1\ldots v_k\psi$ or $\forall v_1\ldots v_k\exists u\psi$ where ψ is quantifier-free. Then $V_\Omega|\mathrm{Sub}(\varphi)$ is prim. rec.

Lemma 1 is routine. As for lemma 2, if φ is (say) $\forall v_1\ldots v_k\exists u\psi(v_1,\ldots,v_k,u)$ and true then all subsentences which are not quantifier-free are of the form $\forall v_{i+1}\ldots v_k\exists u\psi(d_1,\ldots,d_i,v_{i+1},\ldots,v_k,u)$ $(0 \leq i \leq k;\ d_1,\ldots,d_i \in C)$, and also true.

Lemma 3: $V_\Omega|\mathrm{Sub}(\alpha)$ is prim. rec.

Proof: Since $\alpha = \alpha_1\wedge\cdots\wedge\alpha_{25}$, each subsentence of α is a subsentence of some α_i or is of the form $\alpha_1\wedge\cdots\wedge\alpha_j$ $(2 \leq j \leq 25)$. Each α_i is of the form considered in lemma 2, which proves lemma 3.

Thus $V_\Omega|\mathrm{Sub}(\alpha)$ is a prim. rec. full s-val. for α which makes α true.

Now let V be any recursive total valuation and let $\mathfrak{M}$ be the C-model determined by it, that is, $\mathfrak{M} = \langle C, =^{\mathfrak{M}}, \mathrm{zero}^{\mathfrak{M}}, \ldots, <^{\mathfrak{M}}, c_0, c_1, c_2, \ldots\rangle$. We shall derive a contradiction from $V(\alpha) = 2$ (i.e. $\mathfrak{M} \models \alpha$).

For each $n < \omega$, let $\pi_n(v)$ be the formula of L which expresses that v is the nth successor of 0 (i.e. $\pi_0(v)$ is $\mathrm{zero}(v)$ and $\pi_{n+1}(v)$ is $\exists u(\pi_n(u)\wedge \mathrm{suc}(u,v))$). Since $V(\alpha) = 2$ we have $V(\alpha_i) = 2$ for $1 \leq i \leq 25$. In particular $V(\exists x\ \mathrm{zero}(x)) = V(\forall x\exists y\ \mathrm{suc}(x,y)) = 2$. Therefore we can define recursively $d_0, d_1, d_2, \ldots \in C$ as follows: d_0 is the first $c \in C$ for which $V(\mathrm{zero}(c)) = 2$; d_{n+1} is the first $c \in C$ for which $V(\mathrm{suc}(d_n, c)) = 2$.

Since V is recursive, the sequence $\langle d_n | n < \omega \rangle$ is recursive and, by induction on n, it is clear from the construction that $\mathfrak{M} \models \pi_n(d_n)$ for each n.

We now make use of the special properties of the theory N. It is proved in [Sh, Sec. 6.7] that every recursive function is representable in N and it is clear from the proof that the same applies (under a natural notion of representation) to every partial recursive function. Now let A, B be two disjoint recursively enumerable (r.e.) subsets of ω. Applying the representation theorem to the partial function $F(x) = \begin{cases} 1 & x \in A \\ 0 & x \in B \\ \infty & \text{otherwise} \end{cases}$ we conclude that there is a formula $\varphi(v)$ of the language of N such that for every $n < \omega$:

$$n \in A \Longrightarrow N \vdash \varphi(\bar{n})$$

$$n \in B \Longrightarrow N \vdash \neg \varphi(\bar{n})$$

(where $\bar{n}$ is the numeral $S^n 0$).

Transferring this result to our language L we have

<u>Lemma 4</u>: For every two disjoint r.e. sets A, B there is a formula $\varphi(v)$ of L such that for all n:

$$n \in A \Longrightarrow \alpha \vdash \forall v(\pi_n(v) \rightarrow \varphi(v)) \quad ,$$

$$n \in B \Longrightarrow \alpha \vdash \forall v(\pi_n(v) \rightarrow \neg \varphi(v)) \; .$$

Applying this lemma to our model $\mathfrak{M}$, in which α is true, we see that for every two disjoint r.e. sets A, B there is a formula $\varphi(v)$ of L such that for all n,

$$n \in A \Longrightarrow \mathfrak{M} \models \varphi(d_n), \Longrightarrow V(\varphi(d_n)) = 2 \quad ,$$

$$n \in B \Longrightarrow \mathfrak{M} \models \neg \varphi(d_n), \Longrightarrow V(\varphi(d_n)) = 0 \quad .$$

Since V is recursive and the operations $n \mapsto d_n, \mapsto \varphi(d_n)$ are recursive, the set $\{n | V(\varphi(d_n)) = 2\}$ is recursive and separates A from B. But this is a contradiction if we take A, B to be two recursively inseparable disjoint r.e. sets. The contradiction shows that $V(\alpha) \neq 2$, i.e. - α is false under V. This completes the proof of (a).

D. <u>Proof of Part (b)</u>

Let A, B be two disjoint recursively inseparable r.e. sets and let φ be a formula as in lemma 4. Let β be the following sentence -

$$\alpha \vee (\forall x\varphi(x) \wedge \neg \forall x\varphi(x)) .$$

We shall show that β has the properties stated in part (b) of the theorem. The positive statement is easy: Let $V = V_\Omega|\mathrm{Sub}(\alpha)$ be the prim. rec. semi-valuation that makes α true. Let V' be obtained from V by adding β to the domain and letting $V'(\beta) = 2$. Then V' is a prim. rec. s-val. making β true.

Now let V be <u>any</u> s-val. such that $\mathrm{dom}(V) \supseteq \mathrm{Sub}(\beta)$ and $V|\mathrm{Sub}(\beta)$ is recursive. We shall derive a contradiction from the assumption that $V(\beta) = 2$. Without loss of generality $\mathrm{dom}(V) = \mathrm{Sub}(\beta)$.

Assume $V(\beta) = 2$. Clearly $V(\forall x\varphi(x) \wedge \neg \forall x\varphi(x)) = 0$; so $V(\alpha) = 2$, hence $V(\alpha_i) = 2$ for $1 \leq i \leq 25$. Since every atomic sentence is $\mathrm{Sub}(\alpha)$ (e.g. - $\mathrm{sum}(c',c'',c''')$ is a subsentence of $\forall xy\exists z\ \mathrm{sum}(x,y,z)$), V is defined on all atomic sentences, hence V determines a (two-valued) C-model $\mathfrak{M}$ such that $V = V_{\mathfrak{M}}|\mathrm{Sub}(\beta)$ where $V_{\mathfrak{M}}$ is the total valuation function associated with $\mathfrak{M}$.

We can now get a contradiction exactly as in the proof of (a), by producing a recursive separation between A and B: First construct $d_0, d_1, d_2, \ldots$ such that $\mathfrak{M} \models \pi_n(d_n)$ and then conclude that

$$n \in A \Longrightarrow \mathfrak{M} \models \varphi(d_n), \Longrightarrow V(\varphi(d_n)) = 2 ,$$

$$n \in B \Longrightarrow \mathfrak{M} \models \neg \varphi(d_n), \Longrightarrow V(\varphi(d_n)) = 0 .$$

Of course, to get from $\mathfrak{M} \models \varphi(d_n)$ $(\mathfrak{M} \models \neg \varphi(d_n))$ to $V(\varphi(d_n)) = 2$ $(V(\varphi(d_n)) = 0$ resp.) we have to know that $\varphi(d_n)$ is in the domain of V, i.e. - a subformula of β. That is why we put the disjunct $\forall x\varphi(x) \wedge \neg \forall x\varphi(x)$ in β. The proof of the statements in (b) concerning s-val.'s is thus complete.

Now let V be a 3-val. such that $V(\beta) = 2$. We shall show that $V(\varphi) \neq 1$ for every atomic φ. This will imply that $V(\varphi) \neq 1$ for every sentence φ, hence $V|\mathrm{Sub}(\beta)$ cannot be recursive by what we have just proved.

Since $V(\beta) = 2$ and $V(\forall x\varphi(x) \wedge \neg \forall x\varphi(x))$ is (necessarily) 0 or 1, it is clear that $V(\alpha) = 2$, hence $V(\alpha_i) = 2$ for $1 \leq i \leq 25$. It follows easily (by looking at the equality axioms α_1-α_8) that for all $c,d,e \in C$ the following sentences have the value 2 under V:

$$c = c,\ [c = c \wedge c = d] \rightarrow c = d\ ,$$
$$[c = c \wedge \text{zero}(c)] \rightarrow \text{zero}(c)\ ,$$
$$[c = c \wedge d = d \wedge \text{suc}(c,d)] \rightarrow \text{suc}(c,d)$$
$$[c = c \wedge d = d \wedge e = e \wedge \text{sum}(c,d,e)] \rightarrow \text{sum}(c,d,e)$$
$$[c = c \wedge d = d \wedge e = e \wedge \text{prod}(c,d,e)] \rightarrow \text{prod}(c,d,e)$$
$$[c = c \wedge d = d \wedge c < d \rightarrow c] < d\ .$$

Noting that $\rightarrow_3(1,1) = 1 \neq 2$ it follows easily that we cannot have $V(\varphi) = 1$ where φ is of the form $c = d$, zero(c), suc(c,d), sum(c,d,e), prod(c,d,e) or $c < d$. Thus $V(\varphi) \neq 1$ for all atomic sentences φ, which completes the proof.

Remark: The proofs above apply, mutatis mutandis, to the language $L = \{=, \epsilon\}$, if the hereditarily finite sets are taken to be the intended model for α and β. But we do not know an altogether satisfactory analogue of lemma 4 since a rather artificial axiom, referring directly to the canonical enumeration of the hereditarily finite sets, is needed as a conjunct in α (in the proof we know). In other words, α expresses a mixture of set-theoretic and arithmetic properties, not 'natural' properties of the intended (set-theoretic) model.

3. Discussion and Open Problems. (a) As mentioned earlier, our extension of the completeness theorems is of obvious interest if one wants to know about validity in classes of models of restricted (recursion theoretic) complexity; cf. Mostowski [Mo] and the improvement by Vaught [V]. In a sense our extension puts their negative results into proper perspective; specifically, that the set of valid formulae is not arithmetic if the models considered are determined by a satisfaction relation which is recursive when restricted to atomic formulae. By subsection 1 above, we do get an arithmetic set if we consider instead those models which have a recursive satisfaction relation on the full language (since the property of being a Rec-founded tree is obviously arithmetic). In other words, the negative results of Mostowski and Vaught are sensitive to the precise measure of complexity used: to

their particular analysis of what is meant by a 'recursive model'.

(b) As is implicit in subsection 1(e) and at the beginning of subsection 2, our extension is also obviously relevant to establish the stability of results in elementary metamathematics w.r.t. the exact meaning of the abstract notions involved.

(i) In one direction we have stability or 'invariance' of $\{A:T_A$ is $\mathcal{L}$-founded$\}$ since this is equivalent to ordinary logical validity for all $\mathcal{L} \supset \mathcal{L}_0$ provided only that $\mathcal{L}_0$ is a 'basis' for strict Π^1_1 predicates. It is fair to say that this result is not a 'mere' refinement but quite essential for the rôle of 'logical' reasoning in classical mathematics: for most restricted domains of mathematics, we get no more schemata (of the logical language) which are valid for the sets arising in that domain than those that are generally, logically, valid. This is so because, in most branches, all the sets $\in \mathcal{L}_0$ do arise. - Naturally, they don't arise in so-called combinatorial mathematics about finite domains. It is an empirical fact that here there have been few applications of (logical) model theory, and, as a matter of logical theory, validity in finite domains does not admit a complete axiomatization.

The reader familiar with intuitionistic mathematics should note that logical validity seems much less useful there: different schemata are valid for branches which do and for those which do not use choice sequences.

(ii) In another direction, we consider stability of relations between different rules (or, equivalently, validity for different kinds of models) such as the equivalence (**) on page 23 in section 2. So we have the

Question: For which $\mathcal{L}$ do we have

$\{A:A$ has a $\mathcal{L}$-founded derivation without cut$\}$ = $\{A:A$ has a $\mathcal{L}$-founded derivation with cut$\}$

It seems plausible that such $\mathcal{L}$ need not satisfy the condition on $\mathcal{L}_0$ in (i) above, a fortiori $\mathcal{L}$-founded T_A or T_A^{CF} need not be finite (well-founded). For one thing, as we shall see in II.1 on ω-derivations, the analogue, say $(**)^\omega$, to (**) holds under quite weak closure conditions on (the corresponding) $\mathcal{L}^\omega$; in particular, $\mathcal{L}^\omega$-founded ω-derivations need not be well-founded. Also, more generally, cut-elimination does not at all presuppose completeness w.r.t. (usual) logical validity;

for example, it holds for the intuitionistic rules (which are not complete for classical validity).

The question above, together with the exercise at the beginning of I.2, illustrates the kind of use that can be made of $\mathcal{L}$-founded derivations to distinguish between metamathematical results of different 'depth'. The inversion theorems (in I.2) are quite elementary, holding for a general class $\mathcal{L}$; (**) holds for a smaller class (as shown by Stavi's example); and (i) above is even more special.

Some related questions will be put at the end of Section II.

(c) The use of our generalization, from well-founded to $\mathcal{L}$-founded trees, described in (b) should not be thought of as an isolated exercise, but as a typical example of modern mathematical practice. To understand the mathematical behaviour of given operations (here: rules of inference) one extends the domain of definition (here: finite iterations). This can be rewarding (mathematically) even if the origina restricted domain remains the principal area of application of those operations; of course, by (a) and (b) above, our extension is in fact of intrinsic interest.

The general nature of such uses of generalizations is clear enough. If different operations (here: sets of rules) yield the same results in a particular domain, attention to these differences is, by the nature of the case, unrewarding; as one says, they are 'subtle' and (in the original domain of application) boring. At the same time these boring differences must be considered in proofs; thus the theorems stated cannot - as one says - do 'justice' to the methods of proof . More prosaically, in such a situation there is a high chance of errors and omissions. This chance is reduced if one states theorems in 'proper generality', more adequate to the content of the proof; cf. the technique of avoiding errors in convergence proofs of $\Sigma a_n x^n$ for real x by checking the behaviour of $\Sigma a_n z^n$ for complex $z: |z| = |x|$.

Naturally, it is more difficult to be sure about the relevance of our specific choice of generalization, that is, of replacing finite (or, equivalently in the case of binary trees: well founded) by $\mathcal{L}$-founded. For what it is worth our guiding idea is this. To master large finite configurations one simply must *not*[*] use their build-up! – and if one doesn't, the results will automatically apply to suitable infinite configurations. So at least one way of attaining that mastery of large finite configurations is to make a guess at those 'suitable' infinite ones. In our particular extension, the (mathematical) analysis of 'not using the build-up' is: not using operations defined by transfinite recursions from the terminal nodes of the derivations (which are only $\mathcal{I}$-founded), or again: using operations which are *continuous* and depend only on a finite stump of the tree (starting at its root).

It does not seem far-fetched to push the analogy with function theory a little farther, comparing $\mathcal{I}$-founded trees with *analytic data* and rules of inference (given by finite Post systems) with *algebraic data*. Useful algebraic properties can be inferred from analytic behaviour; for example, analytic data about the poles of meromorphic functions in the complex place imply algebraic properties of being a rational function and so forth. So it would be natural to look for $\mathcal{I}$ such that the 'analytic' equivalence: $\{A: A$ has a $\mathcal{I}$-founded R-derivation$\} = \{A: A$ has a $\mathcal{I}$-founded R'-derivation$\}$ *implies* an 'algebraic' or 'intensional' equivalence

namely: R and R' are interconvertible

(by appropriate elementary conversions). There seems to be a good deal of somewhat *inarticulate knowledge* about rules in the proof theoretic literature which could benefit from this type of analysis; for example, the idea of (sets of) rules which are *directly derivable* from each other, and thus in a much closer relation to each other than equivalence w.r.t. finite derivations. At least for *some* relations of this sort it may be *easier to find* an 'extensional' analysis in terms of a suitable choice of $\mathcal{I}$ than to find appropriate conversions between the rules R and R' themselves. Be that as it may, merely remembering the possibility of choosing between various analyses should have a salutary pedagogic effect: one will avoid that peculiar mixture of missionary zeal and hard-sell advertising which advocates of any one method of analysis employ and which repels the reader of logical sensibility.

---

[*] despite the stress that ideological advocates of *finitism* put on the possibility of such a step-by-step build-up (e.g. of the power set of a finite set), on the fact that the configuration *can* be so built up (in principle).

II. ω-LOGIC: TRANSFORMATIONS OF $\mathcal{L}$-FOUNDED DERIVATIONS AND OF ω-REFUTATION TREES

Some of the principal questions below are also of interest when restricted to the case of ordinary predicate logic discussed in I. As already mentioned in I3(b), some matters are easier to settle for ω-logic, in particular, the relations concerning to closure conditions on $\mathcal{L}$ needed for two kinds of equivalences: (i) between $\mathcal{L}$-foundedness and well-foundedness, or between validity in models $\in \mathcal{L}$ and logical validity, (ii) between derivability (on $\mathcal{L}$-founded trees) by use of cuts and without use of cut. However, the single most promising reason for concentrating on ω-logic seems to us to be this:

To state results which express adequately the mathematical content of Gentzen's work on formal arithmetic, one of the true highlights of proof theory. In accordance with I3(c), the extension to trees which are not well-founded, appears to provide concepts adequate for this aim.

Evidently this aim assumes that the usual formulation of Gentzen's work (in particular, his own as a consistency proof of formal arithmetic by means of ϵ_0-induction) is not adequate. The authors of this article do not agree among themselves on the precise degree of inadequacy of the usual formulations. The following exposition represents the view of one of us.

First of all, though realistically speaking Gentzen's metamathematical methods satisfy pretty well (Hilbert's) requirements on mathematical principles that ought to be particularly reliable (on Hilbert's conception of mathematical rigour), the consistency proof simply does not add to our actual conviction of the truth of theorems proved in formal arithmetic. From this angle, the value of Gentzen's proof is essentially negative, in discrediting a certain conception of mathematical rigour.

Secondly, as is well-known Gentzen's formulation is, on the surface, almost ludicrous. The main principle of formal arithmetic is ordinary induction, on ω, and the metamathematical methods use induction on ϵ_0, which is 'bigger' than ω. So clearly some more or less subtle distinction is involved here which is not conveyed by the usual formulation (and, being 'subtle', not likely to be foundationally really satisfactory). These matters will be developed further below and in subsection 1 of the Appendix.

Lastly, quite pragmatically, the role of ϵ_0-induction within Gentzen's proof theoretic analysis is mathematically marginal. It is used in a proof of termination or convergence of a certain transformation procedure of derivations. Would it not be strange to formulate a solution of, say, a differential equation merely by reference to the methods which are used to prove the convergence (of some procedure of approximation)? At best one might hope to find that the method of proof (for example by the convergence test: $|f_{m+1}/f_m| < \rho < 1$) could be developed to show something about the nature of the solution, like analyticity or continuity in the parameters. These further properties are often of obvious significance in the contexts in which the differential equation arises, where the method used in the convergence proof is of no direct interest. The discovery of such additional properties was then the significant step even if those properties are easily derived by inspection of the convergence proof. (What had to be added? Answer: the understanding of the problem which was needed to know what to derive, was not needed to find the convergence proof.)

One of us has made several attempts to formulate results (derived from Gentzen's work) into which ordinals enter explicitly, in particular ϵ_0 in the case of arithmetic; now ordinals are mentioned in the results themselves, not only in the course of proofs. Naturally, these results are best stated for abstract languages (and axioms which are conservative over formal arithmetic) where ordinals arise naturally, which they do not in the context of first order arithmetic. Examples of such results are bounds on Σ_1^1-orderings that can be proved to be well-founded and particularly, bounds (established by Friedman) on the lengths of certain hierarchies which can be proved to exist [Fr]. But it has to be admitted that these applications are somewhat marginal, and in any case they use little of the machinery of Gentzen's analysis; more formally, the applications use only facts about the set of theorems of formal arithmetic, very little about the derivations. Another attempt to discover significance in the ordinals used in Gentzen's work relied on a certain model of the development of mathematical knowledge, by use of socalled autonomous hierarchies [OL]. If indeed these models are faithful to traditional ideas (associated with finitist or predicative doctrines) and if we look closely at those models we simply become skeptical of the significances of those doctrines; cf. the 'negative' uses of Gentzen's consistency proof mentioned earlier. In short, however brave or well-intended these two attempts may have been, we cannot consider them successful.

We are now persuaded to make a third attempt, by use of trees which are not well-founded; naturally, not the ordinals (such as ϵ_0) play a central role here, but certain operations which are defined (not only on well-orderings but also) on $\mathcal{L}$-founded trees, such as exponentiation to the base 2 or 3 or ω (which is defined on any binary relation; and generates well-orderings ($< \epsilon_0$) when applied finitely often to a well-ordering ($< \epsilon_0$)). To make this new attempt plausible let us review quickly the developments which have occurred since the earlier attempts. (i) We have become aware of the intrinsic, model theoretic significance of trees which are ($\mathcal{L}$-founded but) not well-founded, as explained in Section I. At the same time, for such trees restrictions on the (logical complexity of) predicates to which induction is applied are by no means arbitrary; roughly speaking, unless a predicate $P \in \mathcal{L}$, induction on a $\mathcal{L}$-founded tree simply is not valid for P. This is promising because the 'subtle' difference, mentioned on p. 35 between Gentzen's use of ϵ_0-induction and ω-induction applied in formal arithmetic concerns just this matter: the logical complexity of the predicates to which the two principles of induction are applied. (ii) More generally, we have become sensitive to the basic difficulty involved in using proof theory sensibly for model theoretically meaningful results (which concern only the set of provable theorems): there is a high chance of wasted effort since the detailed operations on derivations studied in the (metamathematical) proofs get lost in the statement of the results. This realization was only a first step since the difficulty remained of <u>discovering properties of the derivations themselves</u>, say, structural ones that we want to know about (and apply the methods of proof theory to such matters). The first concrete evidence in this direction, for a sensible use of Gentzen's methods, was obtained by considering (intuitionistic) derivations of existential theorems $\exists xA$ and asking not merely for <u>some</u> term t for which $A[x/t]$ can be formally derived, but for that term t_π which is provided by a given derivation π of $\exists xA$ [Ml].

More substantial evidence is contained in Statman's dissertation on [St], particularly Chapter III.

To combine (i) and (ii) it is natural to go to the extreme where the trees considered are refutation trees (coding counter models $\in \mathcal{L}$) and to extend Gentzen's transformations so that they are well-defined (and of interest!) when applied to such trees. Here the following problem suggests itself (which we have not solved, neither for ordinary nor for ω-logic):

Is there a mapping ρ defined on all trees (regulated by the usual rules) with cut and taking, as values, trees without cut such that, for suitable $\mathcal{L}$,

ρ maps $\mathcal{L}$-founded derivations into $\mathcal{L}$-founded derivations with the same end formula, and

ρ maps refutation trees coding a class $\mathfrak{M}$ of counter valuations to A into a class $\mathfrak{M}^-$ of semi valuations satisfying $\neg A$ where $\mathfrak{M}^-$ is 'contained' in $\mathfrak{M}$; in other words, ρ eliminates not only cuts from derivations but models from refutation trees.

In subsection 1 we give some analogues for ω-logic to the results in Section I, and in subsection 2 we state some results on transformations which were presented by one of us at the Symposium on the Theory of Logical Inferences at Moscow in March 1974.

1. Preliminaries on ω-logic. We shall here confine ourselves to a single sorted language including distinct constants $a_0, a_1, a_2, \ldots$ (for example, 0, s0, ss0,... with a 'successor symbol' $\underline{s}$) and consider models with domain $\{a_i : i \in \omega\}$ (where s is realized by: $a_i \mapsto a_{i+1}$). This is special, in that, generally, ω-logic is applied to (single sorted) languages with a predicate symbol say $\mathbb{C}$ where $\mathbb{C}$ is always realized by $\{a_i : i \in \omega\}$, but the full domain of the models is left open. (A variant of the latter uses many sorted languages and fixes the domain of one of these sorts in all the models.) Even though our results are more 'refined' than the traditional ones (and would therefore be expected to be more sensitive to 'small

differences), the principal conclusions below happen to hold for each of the variants mentioned.

(a) Generalizations of ω-completeness theorems (for various kinds of ω-valuations and various kinds of rules of inference). As in Section I for ordinary predicate logic, for a given formula A, in the (same finite) language of A, we find (primitive recursive) trees T_A^{ω} and $T_A^{\omega CF}$ with the properties:

(i) T_A^{ω} is regulated by the usual rules for ω-consequence with cut and, for $\mathfrak{L}$ satisfying a few recursive closure conditions,

T_A^{ω} is $\mathfrak{L}$-founded if and only if A is true in all (countable) total ω-valuations in $\mathfrak{L}$, which are defined for all formulae in the language of A.

An ω-valuation satisfies of course the 'adequacy conditions' in that it commutes with the propositional operations and $\exists x B$ is valid if and only if $B[x/a_n]$ is valid for some <u>n</u>.

Note that for every formula A of the form $R \longrightarrow B$, where R are the recursion equations of + and x, T_A^{ω} is $\mathfrak{L}$-founded if $\mathfrak{L}$ is contained in the set of all arithmetically definable sets. The reason is that, as is well-known, there is no arithmetically definable truth definition for the (unique) ω-model of R. In short, every arithmetic formula, true or false, has a $\mathfrak{L}$-founded ω-derivation <u>with</u> cut.

(ii) $T_A^{\omega CF}$ is regulated by the usual cut free rules for ω-validity, and for $\mathfrak{L}$ satisfying a few primitive recursive closure conditions:

$T_A^{\omega CF}$ is $\mathfrak{L}$-founded if and only if A is true in all ω-semi-valuations which $\in \mathfrak{L}$ and which are defined on A.

Clearly, in contrast to (i), there are plenty of formulae $R \longrightarrow B$ which have arithmetic ω-<u>semi</u>-valuations of logical complexity Σ_n^0, depending essentially on the complexity of B.

Strictly speaking, as far as completeness is concerned, these results are best possible; in the sense that, for a given A, we have specific trees T_A^{ω}, and $T_A^{\omega CF}$ which are $\mathfrak{L}$-founded ω-derivation trees with, resp. without cuts provided A is valid for total ω-valuations, respectively ω-semivaluations $\in \mathfrak{L}$. Thus validity of A ensures derivability of A by a <u>specific</u> derivation obtained primitive recursively from the formula A itself (only $\mathcal{L}$-foundedness is contingent on validity in $\mathcal{L}$-models.) But as a <u>soundness theorem</u> our result is weak because it asserts validity of A only for those A which are derivable on the specific trees T_A and T_A^{CF}.

<u>Exercise</u>. Analyze the data determining trees $\in \mathcal{L}$, regulated by inference rules for ω-logic, with and without cut, which ensure soundness; that is, if T and T^{CF} are provided with these data, and are $\mathcal{L}$-founded then their end formulae are true in all ω-valuations, resp. in all ω-semivaluations $\in \mathcal{L}$.

<u>Hint</u>. The crucial point is this: when we are given a valuation V in which the end formula is false, the data occurring at the nodes of the trees must determine, say, primitive recursively an infinite path $\pi_V \in \mathcal{L}$ where at the nodes N of π_V all Γ_N are true (in V) and all Δ_N are false. Suppose now that π_V has been defined up to the node N (in T) and that the sequent at N is derived by a cut. To continue π_V one step further it seems necessary to be given the cut formula used: as part of the data at N. (In the case of our canonical trees T_A there was no need to <u>add</u> such information explicitly because it was provided by the specific construction rules for T_A.)

<u>Remark</u>. Experience in many branches of mathematics suggests that it is generally worth <u>proving</u> plausible impressions like the one above concerning the <u>need</u> for extra data at the nodes (besides the formula derived there). Naturally, we can see by inspection that the additional data are used for our <u>particular</u> construction of π_V. The question is whether the soundness theorem is <u>true</u> (for all $\mathcal{L}$-closed under a few primitive recursive operations and for all $\mathcal{L}$-founded trees of formulae, regulated by the rules considered), even if the ω-derivation trees T are not (explicitly) enriched or 'equipped' with the additional data. -- It goes without saying that the <u>distinction</u> involved arises also in the case of finitary derivations, but the <u>difference</u> is less striking. The reader should consider the <u>data</u> determining the trees themselves, whether a tree T is given by its set of initial segments or by a function defined on the nodes N of the universal tree (which is finitely-branching for ordinary logic and infinitely branching for ω-logic) with values determining whether or not N lies on T at all. In the rest of this paper we

shall use the following data:

A derivation tree is given by a function f defined on the universal tree with the root labelled $\langle 0 \rangle$, and nodes at level $n > 0$ being labelled $\langle n, m_1, \ldots, m_n \rangle$ with $m_i \in \omega$. This corresponds to the path choosing the node $\langle 1, m_1 \rangle$ at level 1, its m_2th immediate predecessor at level 2 etc. The value of f at a node N consists either of a sequent $\Gamma_N \vdash F_N$ together with some (conventionally chosen) name of the rule by which $\Gamma_N \vdash F_N$ is inferred, and a location of the premises at the level above N or if our derivation tree does not contain N at all, $f(N) = \emptyset$ or some other symbol indicating that N is empty.

(b) $\mathcal{L}$-founded derivability in ω-logic with and without cut: model theoretic comparison. It follows from (i) and (ii) in (a) that if $\mathcal{L}$ is contained in the class of all arithmetic sets, there are (false) arithmetic formulae A which have a $\mathcal{L}$-founded ω-derivation, with cut, e.g. T_A^{ω}, but none without cut. This should be contrasted with I.1(e) where much weaker closure conditions on $\mathcal{L}$ are enough to ensure equivalence of rules with and without cut for $\mathcal{L}$-founded derivations of ordinary predicate logic.

On the other hand it is quite easy to write down conditions on $\mathcal{L}$ which are 'weak,' in a sense to be described in a moment, and which ensure that

> each (countable) semi-ω-valuation $\in \mathcal{L}$ can simply be extended to an ω-valuation $\in \mathcal{L}$ (preserving of course the truth valuations of the formulae on which the semivaluation is defined).

One such condition is this (besides the usual elementary closure conditions corresponding to syntactic operations):

> If $X \in \mathcal{L}$ so does the ωth jump, say $X^{(\omega)}$.

The reason is clear (and familiar from the theory of ω-models of say analysis, where one begins with an enumeration of all the sets in the ω-model or, alternatively, with the satisfaction relation for the atomic formulae $a_i \in c$ where c are the constants, of the second sort, for sets). To be precise we consider ω-semi-valuations for those formulae A which contain enough information to ensure that: if the semi-valuation is defined on A it is defined on all atomic formulae in the language of A.* Given such a semi-valuation X and any formula B in the language of A, the unique extension to B of the (total) valuation on the atomic formulae is arithmetic in X and so the satisfaction relation for the whole language is certainly primitive recursive in $X^{(\omega)}$.

The closure condition on $\mathcal{L}: X \mapsto X^{(\omega)}$, is certainly weak in the sense that it does not imply

$$\mathcal{L}\text{-foundedness} \Longrightarrow \text{well-foundedness};$$

it obviously does not do so for arbitrary recursive trees; for e.g. $\mathcal{L} = H_{\omega^\omega}$ satisfies the closure condition, but there are recursive trees which are not well-founded, yet have no descending hyperarithmetic path. It remains to verify that the same applies to our particular derivation trees (which are regulated by rules of ω-inference). To do so, consider T_A^ω where A expresses that a certain prim. rec. ordering is well-founded (which can be done in the language of rings with a free function or relation variable) and choose an ordering which is HYP-founded but not well-founded. Putting it differently, the condition: $X \mapsto X^{(\omega)}$ is also weak in the sense that

validity in ω-models $\in \mathcal{L}$ does not imply logical validity in ω-model.

Discussion. In I.3(b) we referred to the results above in connection with a 'corresponding' open problem for ordinary predicate logic. It seems quite clear that more refined analysis is needed for the open problem. The weakness of the closure condition: $X \mapsto X^{(\omega)}$ is patently connected with the restriction to finite formulae in the language of ω-logic. By now it is--or should be--superfluous to remind the reader that this is a mere fragment of the language, or, more precisely, of the class of languages appropriate to ω-logic. There is an imbalance in the syntax itself (where we have finite formulae but infinite well-founded derivation figures) and trivial counter examples (for example, to the interpolation theorem,

---

* Cf. Stavi's use of equality axioms in I.2 which ensure that every ternary valuation is binary on the atomic formulae.

which holds of course for a sensible choice of language*). It might be profitable to refine first the results above on ω-logic for suitable infinitely long formulae and infinite derivations, perhaps in terms of the segment H_Γ of the hyperarithmetic hierarchy where Γ is the length of certain autonomous progressions (introduced by one of us when infected by traditional interest in 'predicative' mathematics and) studied by Schütte and Feferman. Here we know that (*) if the formula A and some well-founded ω-derivation of A by use of cut are in H_Γ so is some well-founded ω-derivation of A without cut. True, we want more; not a result about well-founded derivations but something about, say, H_Γ-founded ω-derivations. But inspection of one of the more careful metamathematical proofs (which does not apply induction on the given derivation to unsuitably complicated formulae), may solve our problem too.

(c) To complete these preliminaries concerning ω-logic with and without cut, we consider, for illustration, a pair of more simply related sets of rules; 'simply' in the sense that it is easy to show that they generate the same sets of theorems on $\mathfrak{L}$-founded trees, for example if $\mathfrak{L}$ is closed under recursive operations. We have chosen these rules because their so to speak model theoretic relations are easily decided, and so--it seems to us--they are, pedagogically, appropriate for illustrating <u>refinements</u> concerning non-model theoretic matters; in particular,

recursion theoretic properties of transformations
of ω-derivations regulated by the rules considered.

<u>Remark</u> (concerning matters of pedagogy). Realistically speaking the recursion theoretic questions of <u>transformations</u> (of derivations) arise only when the model theoretic inclusions (between sets of theorems) are known to hold. In this sense, model theory comes first and recursion theory afterwards. This <u>logical</u> relationship is opposite to the historical sequence of events since people had been working on cut elimination <u>procedures</u> for 25 years before semivaluations were even

*The 'counterexample' is almost as trivial as that for ordinary predicate logic when the constants $\top$ or $\bot$ are left out, $A \to B$ holds, but A and B have no symbols in common.

mentioned (in [Sch]). The reader interested in 'sociological' aspects of logical research may wish to use these facts to reconsider the notorious lack of interest, of the silent majority (of logicians), in proof theory: Was it due to 'prejudice' or to an objective lack of interest? of the results actually stated in the (bulk of) proof theoretic literature.

Repetition. Given a set $\mathcal{R}$ of (formal) rules, let $\mathcal{R}^+$ be the system obtained by adding the rule: from $\Gamma \vdash \Delta$ derive $\Gamma \vdash \Delta$ with the stipulation that the derived $\Gamma \vdash \Delta$ is equipped with (some name of) the rule above, for example, Rep.

Evidently $\mathcal{R}$ and $\mathcal{R}^+$ generate the same sets of theorems on well-founded derivation trees. This (extensional) equivalence obviously holds for $\mathcal{L}$-founded trees too provided $\mathcal{L}$ contains all sets which are recursive in the trees considered. For suppose an $\mathcal{R}^+$-derivation of, say, A is $\mathcal{L}$-founded. It cannot contain an infinite (uninterrupted) sequence of application Rep. dominated by the node N_0 say because, by the stipulation above, such a sequence has no branching and is patently (primitive) recursive in the given derivation trees. So the $\mathcal{R}^+$-derivation collapses to an $\mathcal{R}$-derivation by omitting the rule Rep altogether. The collapsing process is not total recursive, by pp.46-47; it is partial recursive by searching at each N_0 for the last application of Rep in the sequence beyond N_0 (and the search terminates for our $\mathcal{L}$-founded trees).

One important use of Rep, in work considered in II.2, depends on the following result where the trees considered are regulated by $\mathcal{R}^+$ and given by the data f at the end of II.1(a).

There is a (Kalmar) elementary operation, call it Π such that if T is an $\mathcal{R}^+$-derivation then $\Pi T = T$ and, for all T

$\Pi(T)$ is an $\mathcal{R}^+$-derivation (but not necessarily well-founded).

The definition of Π is clear. If $f(\langle 0 \rangle)$ is not well formed, take $\Pi(f)$ to be a derivation of, say, $0 = 0$.

If $(\Pi f)(N')$ has been defined for N' on the path $\langle 0 \rangle$ to $\langle n, m_1, \dots, m_n \rangle$ and $= f(N')$

$$(\Pi f)(N) = f(N)$$

for $N = \langle n+1, m_1, \dots, m_n, m_{n+1} \rangle$ provided $f(N)$ satisfies the data at

$f(\langle n, m_1, \ldots, m_n \rangle)$; otherwise change $f(N)$ so as to satisfy the data and apply Rep beyond N.

Naturally, in general, Πf will not be well-founded.

This simple construction indicates clearly how, with the help of (Rep and) Π, one can extend operations defined on say α-founded derivations to arbitrary trees.

Note, for reference below, that the logical complexity of

$$f \text{ is locally correct, } LC(f) \text{ for short,}$$

that is, f is regulated by our rules, is Π^0_1. Naturally

$$LC(\Pi f)$$

is identically true, but en revanche

$$\Pi f \equiv f$$

is again Π^0_1.

To get familiar with the properties of Rep the reader may consider the following:

Exercise. Find rules $\mathcal{R}$ such that A has a primitive recursive $\mathcal{R}^+$-derivation, but no primitive recursive $\mathcal{R}$-derivation. Hint: Consider a polynomial $P(\vec{x},\vec{y})$, in several variables $\vec{x}$, $\vec{y}$, such that $\forall\vec{x}\ \exists\vec{y}\ P$ is true, but there is no primitive recursive function φ satisfying $\forall\vec{x}\ P(\vec{x}, \varphi\vec{x})$. Use rules $\mathcal{R}$ which allow the inference $\vdash \exists\vec{y}P$ from some $\vdash P[\vec{y}/\vec{m}]$ but (in contrast to the 'usual' rules) not the multiple-succedent inference $\Gamma \vdash \exists y\ P, \Delta$ from $\Gamma \vdash P[\vec{y}/\vec{m}], \exists yP, \Delta$. Then, if $\exists\vec{y}\ P[\vec{x}/\vec{n}]$ is true, say $P[\vec{x}/\vec{n}, \vec{y}/\vec{m}]$, there is a primitive recursive $\mathcal{R}^+$-derivation of $\vdash \exists\vec{y}\ P[\vec{x}/\vec{n}]$ by repeating this sequent m_0 times, where $\max \vec{m} \leq m_0$, followed by a computation of $P[\vec{x}/\vec{n}, \vec{y}/\vec{m}]$. The verification of this state of affairs is primitive recursive.

Put differently, the rule Rep acts as a delay device (in the sense of delaying decisions). This kind of thing is very familiar in recursion theory when r.e. or partial recursive objects are replaced by total or primitive recursive (or even

Kalmar elementary) ones ; cf. the theory of well-orderings or of neighbourhood functions of countable functionals. The passage from partial to total operators raises immediately another problem (which is familiar in the whole of functional analysis).

Extensions of operations beyond their (principal) domain of definition. Even if we are principally interested only in the domain of $\mathcal{L}$-founded derivations or ω-derivations, and consider transformations, say ρ, defined on this domain, the question arises whether ρ can be extended beyond this domain--not trivially, that is, not simply by giving ρ a conventional value outside the domain. Non-trivial requirements on ρ may involve definability properties; for example, whether ρ is the restriction of some primitive recursive functional. Clearly, this particular requirement implies (the weaker requirement of) continuity since (primitive) recursive operations are continuous. (Even the latter is not trivial since, for general $\mathcal{L}$, the domain of $\mathcal{L}$-founded trees is not closed in the space of all trees). In view of the relationships mentioned, one will formulate—when possible--'positive' results asserting the possibility of extensions for (primitive) recursive operations, and 'negative' results for continuous operations.

The matter of continuity or uniformity applies strikingly to the little argument earlier on showing the equivalence between $\mathcal{L}$-founded $\mathcal{R}^+$-derivations and $\mathcal{R}$-derivations (when $\mathcal{L}$ is sufficiently rich). There were 2 cases to consider: whether the $\mathcal{R}^+$-derivation did or did not contain an infinite path consisting only of repetitions (beyond a certain node). Clearly, the decision between these 2 cases cannot be made continuously in the trees considered.

Exercise. Show that there is no recursive (and hence no recursively continuous) mapping from $\mathcal{R}^+$-derivations to $\mathcal{R}$-derivations, for suitable $\mathcal{R}$. Hint: Consider $\mathcal{R}^+$-derivations of formulae $\exists \vec{y}\, P$ for polynomials P with integral coefficients and use the fact that the set of $\mathcal{R}$-derivable formulae $\exists \vec{y}\, P$ is not recursive.*

*Presumably, by use of formulae containing relation or function symbols (in place of $\exists \vec{y}\, P$) one will be able to show that there are no such continuous mappings, even if they are not required to be recursive.

Warning: The problem is to show that there is no recursive mapping from $\mathcal{R}^+$-derivations, to $\mathcal{R}$-derivations; it is evident that the particular 'collapsing' mapping in our little argument described above, is not recursive.

Remark on the transformation of refutation trees. Though we do not go into this matter in the present paper, it is perhaps worth mentioning that the extension of familiar cut elimination procedures to trees which are not $\mathcal{L}$-founded is essential if they are to be applied to refutation trees.

We believe that these preliminaries pinpoint sufficiently the issues that need attention to make the formal details in subsection 2 below almost routine.

2. Cut-elimination: analyzing the issues involved. The reader is assumed to know one of the standard treatments, at least for ordinary predicate calculus, for example by Gentzen or in the books of Kleene, Schütte or Shoenfield. The statements of their main results do not mention explicitly any particular procedure for eliminating cuts, but only the possibility of doing so: the cut rule is said to be permissible (that is, conservative w.r.t. the remaining rules). The standard treatments proceed, generally, by induction on the socalled cut-degree, that is, the logical complexity of the cut formulae used, and on the 'size' of the derivations (which is a finite or countable ordinal in ordinary, resp. ω-logic). The logical complexity of the predicate to which induction is applied, is not analyzed. Thus, though it is clear that the results must hold for suitable $\mathcal{L}$-founded trees which are not well-founded, further analysis is needed to find such $\mathcal{L}$. More to the point, different parts (lemmas) of the standard treatments require different closure conditions on $\mathcal{L}$. Those auxiliary 'lemmas' are liable to have independent uses since we are in fact interested in different $\mathcal{L}$ --at one extreme, in refutation trees (which need not be $\mathcal{L}$-founded except for quite trivial $\mathcal{L}$). To analyze this we need concepts in terms of which these uses' and the relevance of the lemmas can be made explicit. As usual, this last step will need few formal changes in the proofs; it needs an understanding of what we are talking about and what we want. Below the following issues will be treated. (a) The explicit definition of a cut-'reducing' operation ρ on trees T in the sense that (i) if T contains no

cuts, $\rho T = T$ and (ii) if $\underline{c}$ is a bound on all the cut degrees of T and b_N is the degree of any cut formula of ρT at the node N, then $1 + b_N \leq c$ (and so $b_N < c$ if $\underline{c}$ is finite). By use of the operation Π in II.1(c), ρ is extended to <u>all</u> trees T so as to satisfy (i) and (ii) if T is regulated by the rules considered (including <u>Rep</u>). The <u>sizes</u> of T and ρT are compared <u>quantitatively</u> in terms of ordinal functions when T and ρT are well-founded, and generally by means of appropriate operations such as exponentiation to the bases 2, 3 or ω (which are defined for arbitrary relations and satisfy the familiar laws). (b) Given ρ, one asks for <u>closure conditions</u> on $\mathcal{L}$ such that ρT is $\mathcal{L}$-founded, if T is $\mathcal{L}$-founded; for trees T of bounded cut rank, say c, these closure conditions induce new, possibly stricter conditions ensuring that $\rho^c T$ is $\mathcal{L}$-founded, if T is. Lower bounds on these conditions are set by the 'model theoretic' requirement that rules with and without cut generate the same set of theorems by use of $\mathcal{L}$-founded tress; cf. II.1b (we do not discuss the case of infinite $\underline{c}$ in the present paper). The two principal <u>closure conditions</u> so far considered are formulated in terms of comprehension principles satisfied by $\mathcal{L}$ and in terms of socalled bar (=tree) induction or of closure under the corresponding (generalized) ordinal functions. The choice of these conditions is made for formal reasons concerned with the <u>metamathematical principles</u> needed to <u>prove</u> that ρ preserves well-foundedness.

(c) From a more explicit or 'mathematical' point of view, it is not satisfactory to 'hide' the issues involved in restrictions on methods of proof: one lets 'it all hang out'. Specifically, the trees T, given in the first place by the <u>data</u> mentioned at the end of II.1(a), are <u>enriched</u> by adding additional information at each node N; for example by bounds on the 2 measures mentioned in (a), that is, on the size, say σ_N, of the tree T_N dominated by N and c_N, on the cut degrees of all the inferences occurring on T_N. Other enrichments, by adding various (finite) <u>formal derivations</u>, are suggested by p. 122 of [SPT II] and [LE]. We give here just two samples of questions about such enrichments: (i) How can ρ of (a) which is defined on the (fairly) bare trees, be modified to say ρ^+, to operate on enriched trees? Can this be done 'extensionally' in the sense that, if T_1^+ and T_2^+ differ only in the enrichments (and so are the <u>same</u> tree T in the sense of the original data) then so do ρ^+T^+, and $\rho^+T_2^+$; and ρT provides the bare data of $\rho^+T_i^+$ $(i = 1, 2)$. Inspection of standard treatments provides a suitable ρ^+. (ii) How exactly are enrichments related to restrictions on metamathematical principles? Naturally, there are two kinds of restriction: one on the _expressive power_ of the language used (in metamathematics), the other on the _proof theoretic strength_ of the principles (stated in a given language). As to the first restriction, in particular in connection with (b), _something like the enrichment by ordinal_ (notations for the) _bounds_ σ_N above _is needed to paraphrase the notion of well foundedness if the language does not contain functions or set variables_, for example first order arithmetic. Even if free function variables are included, (the hypothesis of) well-foundedness, in (b), cannot be expressed; but _modulo_ a system of notations (defined arithmetically), the hypothesis is replaced by the--generally stronger--requirement that the _tree ordering of the nodes of_ T^+ _be compatible with the given ordering of the notations_ (at the nodes; and the conclusion, that ρ^+T^+ is well-founded, must then be strengthened accordingly). If the notation system and its ordering are defined recursively, the compatibility requirement is Π_1^0. This reduction is, however, still inadequate if the metamathematical language is _quantifier-free_, in which case an implication with a Π_1^0 hypothesis cannot be expressed directly. Additional enrichments, by use of suitable formal derivations, are used for a further reduction because they permit a quite elementary kind of _self correcting operation_ δ, to be compared to Π in III.1(c): if T^+ satisfies the compatibility condition, then $\delta T^+ = T^+$; and, even if T^+ does not, δT^+ does, where T^+ and δT^+ have the same end formula.--As to the use of enrichments for the second purpose, of reducing the proof theoretic strength of the metamathematical principles used, it appears that, at least up to now, this matter always looks after itself: when we demand _explicit_ solutions stated in a language of weak expressive power, the _natural_ metamathematical arguments tend to be quite elementary; cf. the possibility of 'algebraic' treatments even of ordinary number theory if the results are stated in 'suitably' explicit form.

At the symposium on the theory of logical inference (March 1974 at Moscow), one of us presented a particular system of ω-logic, and a treatment of cut-elimination.

The latter contains most of the formal details needed to analyze, in that particular case, the issues (a)-(c) above. We do not repeat the details here, for 2 reasons. The first is pedagogic. Realistically speaking, most readers will be familiar with some standard treatment. So they will not only find a repetition of familiar details boring, but are liable to miss the few novel points altogether. Since the issues in question are quite general, it should be possible to present them in general terms, with an occasional reference to some general features (illustrated by any) of the standard treatments. This is precisely what we shall do. The second reason is, quite simply, this: The details in question are inadequate; literally so for the type of applications of cut-elimination which seem most promising at the present time. Specifically, the choice of language is not properly considered; more seriously, the choice of 'trivial' axioms for closed atomic formulae, that is, true numerical equations, excludes even a possibility of formulating some of these applications. We shall go into these inadequacies in II.2(d) below. Here it should be remarked that these inadequacies do not touch the--unstated--aim of the symposium paper , which accepts the standards of traditional proof theory and simply tries to show how much can be done by primitive recursive methods. The trouble with this aim is philosophical, in the popular sense of the word! The whole project of going into the detailed structure of cut elimination procedures is teratological, if those traditional aims are accepted because they concern only principles of proof and their consequences, not the derivations themselves (as we have said many times before).

(a) Let R be one of the familiar rules with cut, R^- without cut, both supplemented (if necessary) by the rule Rep of II.1(a). Let $\mathrm{Der}(f, \Gamma \vdash \Delta)$ mean that the data f, in the sense of II.1(a), determine a tree T regulated by the rules R (or, equivalently, T is locally correct for R) and that the end sequent of T is $\Gamma \vdash \Delta$; in other words, $f(\langle\rangle)$ consists of $\Gamma \vdash \Delta$ and the (name of the) rule used in the last inference. We are concerned with the property

$$(*) \qquad \forall f \, \forall \Gamma \, \forall \Delta [\mathrm{Der}(f, \Gamma \vdash \Delta) \longrightarrow \exists f' \, \mathrm{Der}^-(f', \Gamma \vdash \Delta)],$$

sometimes also called 'permissibility' of cut. We call R^- closed under cut if

$$(**)\quad \forall f'_1\, \forall f'_2\, \forall \Gamma_1\, \forall \Gamma_2\, \forall \Delta_1\, \forall \Delta_2\, \forall C[(\mathrm{Der}^-(f'_1, \Gamma_1 \vdash \Delta_1, C) \wedge \mathrm{Der}^-(f'_2; C, \Gamma_2 \vdash \Delta_2)) \longrightarrow \exists f'_0\, \mathrm{Der}^-(f'_0, \Gamma_1 \cup \Gamma_2 \vdash \Delta_1 \cup \Delta_2)] .$$

A functional ρ_e is called a cut elimination procedure if $\rho_e : f \longmapsto f'$ realizes (*) and a functional σ is called a cut functional if $\sigma:(f'_1, f'_2) \longmapsto f'_0$ realizes (**).

Remarks (i) As is well-known, for finite trees T with decidable rules, a recursive ρ_e is ensured by the truth of (*); simply run through all f' till you get a locally correct cut free tree with end formula $\Gamma \vdash \Delta$. Similarly, if the trees considered are well-founded, ρ_e can be defined from σ by tree recursion (on the ordering of the tree determined by f), also called 'bar recursion'. If the trees are assumed to be only $\mathfrak{L}$-founded, for elementary classes $\mathfrak{L}$ we do not get automatically a 'trivial' solution of (*) by a recursive (functional) ρ_e nor do we get trivially a solution of (*) from one of (**). Since the trivial solutions above are obviously unintended, our scheme of considering $\mathfrak{L}$-founded trees instead of well-founded ones is at least a candidate for an analysis of our intentions here. (ii) On the other hand, in the presence of Rep, (*) is empty without at least a tacit requirement that ρ_e preserves some kind of $\mathfrak{L}$-foundedness; otherwise we can take $\rho_e(f)$ to be the single line consisting of $(\Gamma \vdash \Delta; \mathrm{Rep})$ repeated infinitely often. (iii) Recall from II.1(c) that, in the presence of Rep, we can extend a definition of ρ_e on

$$\{f : \exists \Gamma, \Delta\, \mathrm{Der}(f, \Gamma \vdash \Delta)\} ,$$

to all f by use of the operation Π. So below we shall tacitly assume that f is locally correct.

We wish to indicate in general terms how to extract cut elimination procedures from standard (syntactic) proofs of the permissibility of cut, (*), usually stated only for finite or well-founded derivation trees. As mentioned earlier, we have in mind the older literature which did not derive (*) from (**) nor did it use the recursion theorem for defining ρ_e from a suitable functional equation for

ρ_e.[*] As so often, the early work lends itself to an abstract generalization because, at an early stage, people rarely understand very well the particular objects they (talk about and) intend to study. So they seldom spot special properties of these objects which can be used to shorten proofs, cf. the return to some early 'syntactic' work on formal systems (with Σ_1 sets of axioms) in later work on predicate logic of $L_{\omega_1 \omega}$ after model theoretic 'shortcuts' had been found for $L_{\omega\omega}$.--In fact, even a casual look at the proof theoretic literature suggests that the arguments apply to $\mathfrak{L}$-founded trees. Ever since Gentzen, this literature shows only the end formula and a few of its predecessors, apart from a lot of dots: evidently an argument cannot possibly depend on what is not shown! such as the terminal points of the trees. In formal language, the older arguments begin with the end formula and proceed 'downwards' in the tree ordering--in contrast to beginning with the axioms and proceeding 'upwards' (where inverted commas are used because, on this terminology, the root $\langle\rangle$ of a tree is at the top).

Proofs of (*) involve generally 3 elements; two of them are, obviously, preliminary, the third is the main (induction) step 'reducing' the degrees of cuts, but not necessarily their number. We shall see that this informal difference, between 'preliminary' and 'main' step, corresponds to a precise distinction in terms of closure conditions on $\mathfrak{L}$ when $\mathfrak{L}$-founded trees are considered.

(i) Derivations can be converted into socalled standard form, that is, any variable-binding operation, say at a node N, binds a variable a_N where a_N and $a_{N'}$ are different for distinct N, N'; for example, in the inference, at N, of $\Gamma \vdash \forall x A, \Delta$ from $\Gamma \vdash A[x/a_N], \Delta$ where a_N does not occur in $\Gamma \cup \Delta$. A consequence of this requirement is that, for any term t, any standard derivation of $\Gamma \vdash A[x/a_N], \Delta$ can be converted into a derivation of $\Gamma \vdash A[x/t], \Delta$ by substituting t for a_N throughout. Clearly, this (substitution) operation is defined generally

---

[*] The standard form of the recursion theorem would apply only to notations for recursive f, not to abstract f, for which one uses the socalled relativized recursion theorem. For more essential defects of this kind of solution, see the discussion at the end of II.2(a), p.58.

(roughly, on all trees with countably many nodes) provided only the language contains infinitely many variables. As a psychological experiment the reader should give a proof of this convertibility by induction on the length of derivations, and see how stilted and ritualistic the proof sounds!

(ii) For the usual classical inference rules (apart of course from cut or modus ponens) all the premises are valid if the inferred sequent is valid; for example as in the Exercise on p. 22, if

$\Gamma \vdash A \wedge B, \Delta$ is valid so are both $\Gamma \vdash A, \Delta$ and $\Gamma \vdash B, \Delta$.

(of course, as shown by cut, this invertibility property is not needed for the soundness of an inference schema). For proof theory, something stronger, an inversion procedure, is of principal interest. For given rules $\mathcal{R}$ the procedure converts an $\mathcal{R}$-derivation of the conclusion of a rule into $\mathcal{R}$-derivations of its premises.

-NB. Even if $\mathcal{R}$ is sound, such a procedure is not ensured by the invertibility, which refers only to validity; trivially, if the rules are not (known to be) complete for validity, but also because the reference to 'convertibility' presupposes a mapping in some restricted class of operations (from $\mathcal{R}$-derivations to $\mathcal{R}$-derivations).

For rules $\mathcal{R}$ with cut (with or without Rep) it is clear that we have (elementary) inversion procedures; in the case above, just cut an $\mathcal{R}$-derivation d of $\Gamma \vdash A \wedge B, \Delta$ with some finite derivation of $A \wedge B \vdash A$, resp. $A \wedge B \vdash B$. The resulting derivation is $\mathcal{L}$-founded for the same class $\mathcal{L}$ as d.--For rules $\mathcal{R}^-$ without cut, any (standard) derivation d^- of $\Gamma \vdash A \wedge B, \Delta$ is treated as follows: The formula $A \wedge B$ is replaced by A, resp. B at each node N, provided $A \wedge B$ is not the argument* of any inference rule applied at the nodes N' on the path from $\langle 0 \rangle$ to N. If $A \wedge B$ is the argument of the inference at N', then the subderivation $d^-_{N'}$ (dominated by N') has the form

*We use the following terminology which is justified by the structure of the rules considered. There is a single formula, say C, either in the left sequent ($\Gamma', C \vdash \Delta'$) or the right ($\Gamma' \vdash \Delta', C$), such that the premises of the rule have the form Γ', $\Gamma_i(C) \vdash \Delta'$, $\Delta_i(C)$ where $\Gamma_i(C)$, $\Delta_i(C)$ depend only on the argument C, not on the parameters Γ', Δ'; arguments are sometimes called 'principal' or 'active' formulae.

$$\frac{\begin{array}{cc} d_1^- & d_2^- \\ \Gamma' \vdash A, \Delta'; r_1 & \Gamma' \vdash B, \Delta'; r_2 \end{array}}{\Gamma' \vdash A \wedge B, \Delta'; \vdash_\wedge}$$

(where r_1 and r_2 are the names of the rules loc. cit., and '$\vdash_\wedge$' of the rule applied at the node N'; the names form part of the data determining $d_{N'}^-$). Our 'inverted' derivation, dominated at N', is then

$$\begin{array}{c} d_1^- \\ \Gamma' \vdash A, \Delta'; r_1 \\ \hline \Gamma' \vdash A, \Delta'; \underline{\text{Rep}} \end{array}$$

This new derivation is locally correct for $\mathcal{R}^-$ since by hypothesis the formula $A \wedge B$ is a parameter of the inferences between $\langle\ \rangle$ and N'.--Our 'treatment' of d^- is evidently (primitive recursively) continuous since, roughly speaking, we need look only at the finite segment $\langle\ \rangle$, N of d^- to determine the data at the node N of the 'inverted' derivation. (We say 'roughly speaking', because we have left open just which data are used to determine the trees themselves, in the sense of the Exercise in II.1(a) above.) Once again, $\triangleleft$-foundedness is clearly preserved. Of course, we need not decide at N whether or not $A \wedge B$ will 'ever' be the argument of $\vdash_\wedge$ anywhere in d_N^- (unless '$\vdash_\wedge$' occurs at N itself). If such a decision were necessary, the operation of inversion could not be expected to be continuous.

Pedagogic Remark: The discussion above shows--beyond any shadow of doubt--that it is possible to present familiar proof theoretic arguments in relaxed detail, an obvious prerequisite for inspiring some degree of confidence in the reader; after all, if the author is not interested enough in a topic to give it detailed attention, what reason is there to suppose that the matter is, objectively, worth attention? In connection with our present topic of cut elimination , the views of the authors vary between two extremes; one finds that the introduction to II.2 is ample and that II.2(a) - II.2(c) are not justified without some striking application, such as solving the open problems mentioned near the end of II.2(d); the other extreme would have us give a detailed treatment of the main induction step too (perhaps for some specific system of rules). Reluctantly, we decided on the compromise below.

(iii) After these preliminaries, we 'consider' the main step in the standard proofs of (*), the permissibility of cut. More formally, we set up a <u>functional equation</u> (for ρ) which is evidently satisfied by a <u>primitive recursive</u> operation. At one step the rule <u>Rep</u> is used--apparently in an essential way. Here the reader should recall the Exercise in I.1(c) on the role of <u>Rep</u> in replacing partial recursive by primitive recursive operations.

The aim is to make ρ cut 'reducing', that is, the maximum degree is reduced by ≥ 1. This is <u>prima facie</u> a global property of a derivation (even) for derivations of finitely bounded cut degree. However, the <u>continuity</u> properties of ρ ensure that the property follows from <u>local conditions</u>, roughly as follows. What happens at the node N of ρT is determined by what happens on a finite number of segments of T in the <u>stump</u> of T between $\langle\rangle$ and (the level of) N. Now ρ is such that if in this stump (of T) there is no cut of degree $> n+1$ then there is no cut of degree $> n$ in this stump of ρT: cuts are pushed 'down' the tree ordering. If the cut degree of T is finite and $= n+1$, and if ρT contained a cut of degree $c > n$, there would be a node N of ρT with cut degree c. But this contradicts what we just said. Put more positively, if ρT has a cut of degree n at N then T has a cut of degree $n+1$ in the <u>stump</u> between $\langle\rangle$ and N.

To avoid misunderstanding: when considering the main step or, as we said, more formally: when formulating various clauses in the <u>functional equation</u> for ρ--we give, in effect, a uniform scheme for reducing 'simultaneously' infinitely many cuts. Of course when a cut, say a maximal one in the tree ordering of a given derivation d has been 'considered' and replaced by cuts of lower degrees, the latter are not 'reconsidered' in this simultaneous reduction; we go on 'down' the tree <u>d</u>.

The standard treatment applies straightforwardly to the cases where the cut formula C at N is the argument of <u>both</u> rules (r_1 and r_2) at the immediate predecessors N_1 and N_2. For example, if C is $\neg C'$ and $\vdash_{\neg C'}$ is the name of the cut rule

$$\frac{\dfrac{\dfrac{d_1}{\Gamma_1, C' \vdash \Delta_1;\ r_1'}}{\Gamma_1 \neg C', \Delta_1;\ \vdash\neg} \qquad \dfrac{\dfrac{d_2}{\Gamma_2 \vdash C', \Delta_2;\ r_2'}}{\Gamma_2, \neg C' \vdash \Delta_2;\ \neg\vdash}}{\Gamma_1 \cup \Gamma_2 \vdash \Delta_1 \cup \Delta_2;\ \vdash_{\neg C'}}$$

then $\rho(d_N)$ is given, for suitable r_1'', r_2'', by

$$\frac{\dfrac{\dfrac{\rho(d_1)}{\Gamma_1, C' \vdash \Delta_1;\ r_1''} \qquad \dfrac{\rho(d_2)}{\Gamma_2 \vdash C', \Delta_2;\ r_2''}}{\Gamma_1 \cup \Gamma_2 \vdash \Delta_1 \cup \Delta_2;\ \vdash_{C'}}}{\Gamma_1 \cup \Gamma_2 \vdash \Delta_1 \cup \Delta_2;\ \underline{\text{Rep}}}$$

The new cut formula C' is patently of lower degree than the cut formula C, that is, $\neg C'$ which is eliminated.

Remarks. (i) The 'suitable' r_1'', r_2'' are either r_1', r_2' or Rep, depending on the end piece of d_N. (ii) The use of Rep, in the last inference of $\rho(d_N)$ is required to make sure that any stump of ρT is determined by the corresponding stump of T, as required above. Put differently--in more civilized language--a modulus of continuity of ρ at the argument $(\langle\rangle, N)$ should be bounded by the level of N.--The reader should verify that, by help of this property, the action of ρ, that is, the value of $(\rho T)(N)$ is nicely expressed in terms of values $T(N_i)$ for (a finite number of) nodes N_i at levels between $\langle\rangle$ and the level of N. The positions of the nodes N_i depend of course not only on the position of N (in the universal tree), but also on the formulae and names of rules occurring at the N_i, that is on the values $T(N_i)$.

If C is a conjunction, we have two cuts. If C is a universal formula, say $\forall x A$, and $\Gamma_1 \vdash \forall x A, \Delta_1$ has the derivation

$$\frac{\dfrac{d_1}{\Gamma_1 \vdash A[x/a_{N_1}], \Delta_1;\ r_1}}{\Gamma_1 \vdash \forall x A, \Delta_1;\ \vdash\forall}$$

we convert d_1 into standard form, and call it d_1 too. Hence we have derivations $d_1[a_{N_1}/t]$ for $\Gamma_1 \vdash A[x/t], \Delta_1$ where, by definition, the degree of $A[x/t]$ is less than that of $\forall x A$ (however 'complicated' t may be). Now suppose $\Gamma_2, \forall x A \vdash \Delta_2$ has the derivation d_2. Whenever (an ancestral occurrence of) $\forall x A$ is the argument of an inference, that is, of $\forall\vdash$, in d_2, this consists in contracting $A[x/t], \forall x A$ to $\forall x A$, for some term t. These uses of $\forall x A$ are suppressed altogether, and replaced by cuts with $d_1[a_{N_1}/t]$ (the cut formula $A[x/t]$ having lower degree than $\forall x A$--as in the standard treatment). Thus, if the rule $\forall\vdash$ is applied in d_N at some node N', inspection of the finite segment $\langle\rangle, N'$ determines whether there is some node N 'above' N' with the cut formula $\forall x A$, and thus tells us what to do at N'. Note that new derivations are added in this passage from T to ρT (ρ so to speak 'stretches' T) and hence we have the bound on the modulus of continuity of ρ required in Remark (ii) above (p.56).

Next we have the case where the cut formula at N is not the argument of both rules at the predecessors N_1 and N_2 of N. Standard uses of inversion principles are taken care of by (ii) above. The (only) delicate situation arises when we have that seesaw of cuts involved in what Gentzen called 'end piece ' or the similar figure which, in the related context in the style of natural deduction, is called 'main branch.' In the case of finite derivations these objects are of course finite; but also in the case of well-founded (infinitary) ω-derivations. For general $\mathcal{L}$, they cannot be expected to be finite (they are if $\mathcal{L}$ satisfies König's lemma for finitely branching trees in the sense of the Appendix). Since the 'main' steps in the standard treatment involve the whole end piece, some device is needed to ensure the continuity of ρ. An obvious step is to use Rep as follows:

> Starting at N and going down the tree ordering, we apply Rep k times to the end formula of d_N unless the end piece of d_N has depth $< k$.

Since the end piece is finitary, one can decide primitive recursively (in the data of d_N) whether or not the proviso is fulfilled.

Warning: Even if d_N does not contain an infinite recursive path, ρd_N may contain one; in this case d_N will contain an infinite recursive (finitary) tree, and hence an infinite path $\in \Delta_2^0$ (or even of degree $< 0'$--by familiar basis results).

Before describing the relevance of this warning to topic (b) of the introduction to II.2, we suggest to the reader that he should 'compare'--as one says-- (i) the functional equation for ρ which is here extracted (in outline) from standard proofs of the permissibility of cut, and the primitive recursive solution of this equation with (ii) the functional equations without the use of Rep and their solution by means of the recursion theorem, e.g. in [C]. The suggestion would be quite silly (if not empty) without some indication of the features which are significant for the comparison. First of all, if $\mathcal{J}$-founded trees are considered at all, the simple look of uses of the recursion theorem is deceptive since one must check if the proofs use induction of suitably limited complexity. More importantly, the recursion theorem operates on notations, and so will often yield unique solutions to a functional equation even if the latter possesses infinitely many 'extensionally' different solutions! Thus the work with the recursion theorem is incomplete, unless the structure of those 'unique' solutions is further analyzed.

Digression on applications to infinitary derivations of infinite formulae (not: to the ω-logic of finite formulas), where the recursion theorem was first used in the way mentioned above. We are well aware of the fact that formulae in the finite language of e.g. first order arithmetic can be 'replaced' by infinite propositional formulae which express the same proposition. In particular, as Tait has stressed, for (hackneyed) questions of consistency or underivability of some schemes of transfinite induction, the answers for arithmetic can be read off from the corresponding work for infinitary propositional logic. But this procedure turns out to be a bit of a detour when examined more closely. Given the mapping from (formulae of) ordinary first order arithmetic to infinite propositional formulae, one translates finite derivations with induction into infinite derivations of propositional logic, and then applies cut elimination for infinitary propositional calculus. But to use this work for less hackneyed results, for example for the less obvious metamathematical properties of arithmetic itself (such as delicate reflection principles discussed further in II.2(c) below), the cut-elimination procedure for the infinitary language has to be formalized; this needs finite codes for the infinite formulae. Inspection shows that, when one starts with a finite (quantified) formula (of arithmetic) each infinite formula that occurs in the transformation is equivalent to some finite arithmetic formula! So the most natural codes for these infinite formulae are precisely their finite equivalents. This reservation about the use of infinite formulae for the present quite specific purpose is of course consistent with the Discussion in II.1(b) on the use of infinite formulae for solving other questions.

(b) For which $\mathfrak{X}$ does ρ of (a) preserve $\mathfrak{X}$-foundedness? If the answer is to be formulated in terms of instances of the comprehension principle satisfied by $\mathfrak{X}$, pretty sharp results are known. As to lower bounds, obviously, ρ does not preserve $\mathfrak{X}$-foundedness if some $\Gamma \vdash \Delta$ has a $\mathfrak{X}$-founded derivation of cut degree $n+1$, but not of cut degree n; cf. II.1(b) for (model theoretic) information. As to upper bounds, by the Warning at the end of II.2(a), it is sufficient that $\mathfrak{X}$ satisfy Δ_2^0-comprehension (with parameters) or equivalently Π_n^0-comprehension (for $n = 1, \ldots, \infty$), often called: arithmetic comprehension principle with parameters.* Provided the derivations considered are of bounded cut degree, say c, our ρ^c preserves $\mathfrak{X}$-foundedness for such $\mathfrak{X}$ and is a cut elimination procedure, called ρ_e at the beginning of II.2(a).

For further progress one has to pay attention to the choice of concepts in terms of which sharper results can be stated. But we have not considered the

*Amusingly, by an observation of Friedman mentioned in the Appendix, this is equivalent to the requirement that $\mathfrak{X}$ satisfy König's lemma for semi-infinitely branching trees. But we do not know a proof of the preservation of $\mathfrak{X}$-foundedness, in which this alternative is used.

matter sufficiently to have any serious comments. Naturally the literature has considered many of the obvious questions and obvious parameters in these areas; for example, extensions to systems with infinite formulae and parameters of ordinal length of formulae and ordinal bounds on cut degrees. It is perfectly possible that results in terms of these quite simple minded syntactic parameters will turn out to be useful. But it seems at least equally likely that better results will involve more delicate restrictions on the class of formulae used, involving their (mathematical) content and not only their (logical) form.--We have in mind the following analogue in the area of ordinary predicate logic, in particular, on decidability, where there are striking results for mathematical theories, that is for classes of formulae of the form: $A \longrightarrow F$ (A being the axioms and F any formula in the language of A). These classes are quite artificial for any of the socalled logical classifications, e.g. by means of logical complexity; those classes had to be discovered by reference to their mathematical content.

(c) What can we expect from enrichments? of the kind described in the introduction to II.2. The first possible use that comes to mind is quite general: functional equations can often be solved by operations in a more restricted class of functionals if the arguments (functions) are suitably enriched. By II.2(a), for solving our particular equation (for ρ), no enrichments are needed as long as the restrictions considered involve only continuity or (even) primitive recursiveness (except that, perhaps, the effects of the rule Rep can be achieved by suitable enrichments). As mentioned in the introduction to II.2, even more is true: not only are the enrichments not needed, they do not even 'affect' the solution, in the sense that ρ of II.2(a) can be enriched to ρ^+, in such a way that the relevant retract of ρ^+ is ρ itself.

Exercise. The reader familiar with 'ordinal assignments' to (or, in civilized language, ordinal bounds for the length of) well founded ω-derivations may find the following hints sufficient for working out the details (assuming, more speculatively, that he wants to do this). On the one hand the standard estimates, for the increase in the ordinal length of a derivation by reducing the cut degree (by 1), are independent of the particular rules used in the derivation. Furthermore the properties of the ordinal functions used in these estimates, 2^α, 3^α or ω^α etc., are simply their 'recursion equations' and continuity at limits, which are valid also when applied to order types that are not well-founded; the corresponding operations on trees also apply generally and not only to well-founded trees. A final modification is needed because of the (novel) use of Rep. This is minor since each application of Rep introduces at most an ω-sequence, and hence changes bounds for α to $\omega\alpha$; so if, for example, $\alpha = \omega^\omega$ then $\omega\alpha = \alpha$. In short we have an easy extension to $\mathcal{L}$-founded derivations. But it may fairly be said that the familiar formulation for the class of well-founded trees particularly when restricted to trees of ordinal $< \epsilon_0$) is defective (and possibly misleading; for example, if the formulation introduces the binary function: $\alpha, \beta \mapsto \alpha^\beta$; this is, of course, defined for $\alpha < \epsilon_0$ but looks as if well-foundedness of α were relevant). Naturally, the reader will not forget here that a correction of such defects requires not only some formal restatement but also a selection (among all correct reformulations), in short, a significant restatement. Part I was devoted to establishing the significance of a restatement in terms of $\mathcal{L}$-founded trees.

The significance of $\mathcal{L}$-founded trees may add some interest to a project pursued by one of us--of axiomatizing, for a given measure of complexity c, the class of theorems of a theory, which are of complexity $\leq$ c, by means of axioms of complexity $\leq$ c; for example, in the case of arithmetic by means of α-induction (for $\alpha < \epsilon_0$) restricted to predicates of complexity $\leq$ c ; cf. p. 331 of [SPT] or [P]. Other variants of this project establish reflection principles (for arithmetic) applied to formulae of complexity $\leq$ c using as metamathematical means the principle of ϵ_0-induction applied to predicates of complexity c"; in particular, quantifier-free ϵ_0-induction for the case of quantifier-free formulae. In this connection, enrichments of (infinite) ω-derivations by adding (finite) formal derivations were envisaged, in particular, on p. 122 of [SPT II] and used in [LE], but with different answers to the obviously crucial questions:

What are these formal derivations, say D_N, at the node N to prove?

From which formal system should we take D_N?

Neither [SPT II] nor [LE] seems to us to face these issues squarely. The former chooses quantifier-free systems and has D_N prove that the subtree dominated by N, assumed to be enriched by ordinal notations, is (i) locally correct and (ii) compatible with the tree ordering. (Since both properties are Π_1^0 they can be expressed in free-variable systems.) The purpose of the choice was evidently 'reductive', avoiding the use of logically complex operations in the explanation of the meaning of logically compound formulae.* But this conflicts with the evidence, collected in the present paper, in favor of using abstract language in metamathematics; the choice of [SPT II] is bad since a formal derivation D_N can't possibly prove anything about an abstract tree! The data determining the trees involved must at least be defined in the language of D_N. Of course, the choice 'works' for the specific project mentioned since the particular trees that turn up, are (primitive) recursively defined. But there is nothing terribly exciting about this particular project. One of us would go so far as to say that we probably have a good chance of finding a better project simply by requiring that the choice above be inadequate for solving the problem!

In [LE] (or, more precisely, in a preliminary draft, since we have not seen the final text) there is a different choice: D_N simply proves the sequent at N, and the formal system used is first-order arithmetic itself (with intuitionistic logic, the corresponding results for classical logic having been proved earlier by an ad hoc trick, using properties of the socalled no-counter example-interpretation [KL]). This choice is perfectly meaningful for abstract trees, though this --for us, crucial--fact was of course not used, and not even mentioned in [LE].

---

* This was needed to avoid the defects of socalled operational semantics, discussed loc. cit. This interpretation explicitly aims to be reductive, yet uses logically compound expressions in its explanations! This pointless enterprise is to be contrasted with familiar explanations of the (classical) meaning of logical operations since these explanations are not intended to be reductive.

However, it is not clear to us that this choice is adequate for the project of getting sharp results, for example, of proving the sharp versions of the reflection principle mentioned above, for formulae of bounded complexity.

(d) Second thoughts on the paper presented at the Symposium on the Theory of Logical Inference, Moscow, 1974. As mentioned in the introduction to II.2, we are not altogether satisfied with (i) the choice of language for ω-logic nor (ii) the choice of axioms (and rules) for atomic formulae. It seems necessary to discuss these defects in some detail just because they do not affect the (traditional proof theoretic) aims of that paper; such defects do not spring to the eye of the reader who is prepared to accept an author's aims (and is therefore particularly in need of outside help before he questions traditions).

(i) The language considered is quite familiar from the proof theoretic literature such as Schütte's book [Sch1]: first order arithmetic with function constants (besides + and $\times$) with axioms to be discussed further in (ii) below and free (monadic) function variables. The logical operations are $(\rightarrow, \wedge, \forall, \exists)$, $\neg A$ being defined as $A \rightarrow 0 = 1$ and $A \vee B$ as $\exists x[(x = 0 \rightarrow A) \wedge (x \neq 0 \rightarrow B)]$. The rules of logic are intuitionistic; thus since atomic formulae are decidable, the classical fragment $(\neg, \wedge, \forall)$ is embedded in the set of theorems. The infinitary rules make the system complete* for ω-models as far as the fragment $(\neg, \wedge, \forall)$ is concerned.

To the student of ω-logic the language is artifically restricted. Not even socalled second order arithmetic can be expressed in this language since there are no function quantifiers. For the preoccupation of traditional proof theory this is not essential. For on the one hand, validity in countable ω-models is equivalent to validity in all ω-models; and on the other, the notion of countable ω-model (of a formula of second order arithmetic) is perfectly naturally expressed in the restricted language as follows.

*Since the matter of completeness of the (intuitionistic) system is not directly connected with the present paper we do not go into it here; except to remark that the familiar incompleteness results, for example, on the assumption of Church's thesis, apply to the case of ω-logic too.

A function variable is used to 'code' or enumerate all sets of an arbitrary countable ω-model and quantification over sets in that model is then expressed by numerical quantification (over the codes of the nth set). In this way, the restricted language is quite suitable for standard questions concerning the set of ω-consequences of given axioms in the unrestricted language.

However, for the kind of structural properties that are of interest to us, the 'reductions' or 'embeddings' just described are quite inadequate, and even deceptive. To get an idea of the depth of the deception involved the reader is recommended to look at Chapter I of Statman's dissertation [St], concerning explicit definitions (and the choice of derived rules or 'axioms' which are demonstrably satisfied by the defined expressions). A profound misconception is involved whenever two objects differ trivially in terms of that conception, but differ strikingly in structure and general behaviour (when viewed by the light of nature).

To be specific, it is simply not true that the structure of derivations in classical logic formulated in the 'full' language $(\neg, \wedge, \vee, \rightarrow, \forall, \exists)$ is represented faithfully in the fragment $(\neg, \wedge, \forall)$ in which $(\vee, \rightarrow, \exists)$ can be defined. Nor does the intuitionistic theory for the fragment $(\neg, \wedge, \forall)$ reflect the classical theory (though it contains the same class of theorems) since, obviously, some classical derivations in that fragment are not intuitionistically valid.

More positively, we shall now formulate a classical analogue to the problem solved by the socalled E-theorems by one of us [Mi] for intuitionistic systems, where attention to structural properties is essential. The E-theorems associate to each derivation $\underline{d}$ of an existential theorem, say, $\exists x A$ a term t_d such that $A[x/t_d]$ is (formally) derivable. And t_d is seen, by inspection, to express the definition which the proof (say $\bar{d}$, expressed by) $\underline{d}$ provides (when we 'unwind' $\bar{d}$). As mentioned already, the E-theorems constitute the first genuine application of cut-elimination or normalization: the term t_d actually occurs in the normal form $|d|$ of d, in particular, in $A[x/t_d]$ from which the end formula $\exists x A$ of $|d|$ is inferred. Now, in the case of a classical derivation

d^c of $\exists x A$ there is of course no guarantee that there is any t for which $A[x/t]$ is known to be true, let alone formally derivable (in the system considered); and *a fortiori* there is no guarantee that $\bar{d}^c$ provides such a t.

Problem: To determine for any given derivation d^c of $\exists x A$ (i) whether or *not*: $\bar{d}^c$ provides a term t_d^c which realizes $\exists x A$ and if it does, (ii) whether or not $\bar{d}^c$ provides a formal derivation, say d' of $A[x/t_d^c]$ in the system considered.--Of course, it is not assumed that d' is intuitionistic since, after all, questions of explicit realizations are quite independent of matters of constructive provability.

A simple minded *candidate* for a solution is just this: find suitable rules of normalization or cut elimination: $d^c \longrightarrow |d^c|$ and see if a theorem of the form $A[x/t]$ actually occurs at a node of $|d^c|$. (People not disturbed by the possibility of systematic error may be interested in the purely *formal* problem: Do different current rules of normalization lead to the same t?)

Clearly, it would be premature to assume that the relevant information contained in $\bar{d}^c$ is preserved when d^c is 'translated' into a derivation of $\neg\forall x \neg A^-$ where A^- is the (canonical) equivalent of A in $(\neg, \wedge, \forall)$.

(ii) As observed at the beginning of (i), the symposium paper takes as axioms a set of atomic formulae closed under cut or under the rule of equating equals to equals. (Here there is simply no problem of proving closure under cut for atomic formulae.) The function constants considered are taken to be computable; at least tacitly, we are assumed to have a *valuation* function for terms built up out of those constants. Certainly, the choice of such sets of axioms is perfectly natural for traditional proof theory which assumes that we are really in doubt about the validity of the formal principles of proof studied, but not about the validity of the metamathematical methods to be used. Then a consistency proof by means of such methods, answers a basic question about formal derivations of Π_1^0 sentences $\forall x[f_0(x) = 0]$ where f_0 is some function constant. The consistency proof would tell us that, for each n, $f_0(s^n 0) = 0$ is one of the axioms (in our set) if the Π_1^0 sentence is formally derivable.

But if we have no doubt about the validity of the formal principles (and know that those Π_1^0 sentences are true) we cannot avoid the question:

What more does a formal proof of $\forall x[f_0(x) = 0]$ provide than that $\forall x[f_0(x) = 0]$ is true?

(This is to be contrasted to the case of say a Π_2^0 statement $\forall x \exists y B$ where a formal proof provides a function $f: x \longrightarrow y$ such that $\forall x\ B[y/f(x)]$ is true and f lies in a restricted class depending on the formal principles considered). More specifically,

What more does a constructive proof of $\forall x\ [f_0(x) = 0]$ provide than can be expected from a non-constructive one?

One of us is attracted by the

Conjecture: A constructive proof $\underline{d}$ provides a specific sequence d_n of computations of $f_0(s^n 0) = 0$, and, in general, a non-constructive proof does not.

NB. Contrary to an almost universal (and almost amazingly thoughtless) misconception, a constructive proof does not generally provide a particularly efficient means of computation! After all, when we have a proof of $\forall x\ [f_0(x) = 0]$, one very efficient (and sound) rule of computation is simply this: put $f_0(s^n 0)$ equal to 0. Now, suppose f_0 is such that we have a non-constructive, but no constructive proof of $\forall x\ [f_0(x) = 0]$, e.g. if this formula expresses the consistency of set theory. Then constructive rules are bound to be less efficient computationally.

Clearly, we cannot hope even to begin a study of the Conjecture (by means of cut-elimination) if we use the fact that all true formulae $f_0(s^n 0) = 0$ are among the axioms: we thereby avoid any analysis of (numerical) computations or of their relation to proofs of identities, that is, formulae of the form $\forall x\ [f_0(x) = 0]$. -NB. The conjecture requires a proper choice of normal derivation! For example, it is clear from inspection that many $\underline{d}$ provide computations d_n which use recursion equations for functions that do not occur in the definition of f . Thus a choice which makes all normal derivations of $f_0(s^n 0)$ simply computations from the defining equations for f_0 (as in [ML]) would be quite inappropriate here.

3. Reappraisals. (a) We hope that our (combined) knowledge of proof theory and other logical literature is wide enough for sound judgment; in other words, for informed judgment. Granted this we need have no doubt that our abstract formulations provide a much better basis, for analyzing existing methods and notions in proof theory, than what is done elsewhere (we mean analysis in terms of closure conditions on $\mathcal{L}$, for $\mathcal{L}$-founded trees and various kinds of valuations in $\mathcal{L}$). In particular, we believe that we have tightened the standards of scientific rigour used in current (proof theoretic) practice; its standards are more like those of an immature science than what could reasonably be expected after more than 50 years(of proof theory). We mention 3 examples.

(i) Systems and methods are accepted as 'natural' without any determined attempt to distinguish between a meaningful sense of 'natural' (that is, modes of thought to which we return even after trying out alternatives) and mere familiarity (to a particular author often ignorant of existing alternatives).

(ii) Mere coherence is accepted as evidence for an adequate analysis, for example, an analysis of the significance of formal discoveries of obvious interest (and, as happens sometimes, such an obvious appeal can be obviously reliable even if it is difficult to analyze the nature of the interest). Specifically, the significance of Gentzen's formal discovery of cut elimination or normalization was sought in its role for socalled operational semantics which, in turn, was put to no other test than that of coherence (which we regard at best as a useful negative test). For a long time, no attempt was made to find other properties (of the procedures above) which could account more convincingly for the obvious appeal of Gentzen's discovery. (This has changed, especially with Chapter III of Statman's dissertation [St]).

(iii) Isolated observations, admittedly of some logical charm, are elaborated, often ponderously, without any attempt at distinguishing between mere curiosities and useful scientific tools. This happened, for example, in the proof theory of socalled subsystems where some (unexpected) relations turned up between restrictions on methods of proof and the content of the theorems proved.

But these relations were not tested for their interest, as if people were paralyzed by the surprise of finding such relations at all.

We are quite aware that judgments on the points raised in (i)-(iii) need not be easy. But similar questions arise in (the development of) other sciences too, and they have been satisfactorily settled; often, quite prosaically, by the accumulation of knowledge and pointed critical comparison of alternatives.--For reference in (c) below, a 'peculiarity' of proof theory should be mentioned, a kind of pun: the various results (which we should like to put through tests) were accepted simply for their claimed 'philosophical interest.' But what was meant--and we may agree--was that many of them had philosophical character; we see no guarantee that they are therefore (even: philosophically) interesting.

(b) We certainly do not wish to overstate the interest of the present paper (whatever its virtues compared to current practice may be). In fact, a particularly critical attitude is appropriate--as with any work that takes traditional proof theory as its starting point. We must not forget that this theory has its roots in Hilbert's program, which arises from untenable epistemological hypotheses. We may not like Jehovah's threats to punish the sins of the fathers on the third or fourth generation; but the brutal fact remains that there are correlations between (the tendencies of) related generations. Similarly, in science there are correlations between the qualities of hypotheses and of the work done on them. We do not expect that much, let alone the bulk of work concerned with the properties of the (hypothetical) ether has a permanent place in science --however natural or coherent the idea of the ether may have been to those who dabbled in the subject.

Conversely, we do expect that some facts which people have stumbled on in the pursuit of a false or even absurd hypothesis will be of use provided the interpretation of these facts is rigorously disentangled from the hypothesis. Some readers may (not) like the comparison with the land Columbus stumbled on when he followed his false* hypothesis about a sea route to India along the latitude of Spain (or thereabouts).

As a practical conclusion, to which we return at the end of (d) below, we expect the most interesting developments of existing proof theory to concern topics which are far removed from its original aim, which was to formulate and support a particular (untenable) hypothesis.

(c) To avoid (what would be) a basic misunderstanding, it should be stressed that the views expressed in (a) and (b) are by no means 'anti-philosophical'; in fact, they are not even directed against traditional philosophy--neither in its academic sense of traditional questions (or 'puzzles') nor in its popular sense of philosophical contemplation. We merely advocate certain distinctions.

As far as the heuristic or, more generally, epistemological value of the philosophical tradition is concerned, it seems necessary to distinguish between different stages of knowledge. At one stage (which certainly occurs, both onto-genetically and phylogenetically speaking) we know very little, and have only first impressions. As a matter of historical fact, the traditional philosophical questions present(ed) themselves to us when we know little. Contrary to fashionable doctrines, it may safely be assumed that the questions are meaningful at that stage. We should go further. When we know very little, they may be the only (meaningful) questions to ask! Suppose we know very little about some physical phenomenon, for example, light; not enough to ask specific questions (for example, about its velocity, at a stage when we do not even know if it has a well-defined velocity or how to measure it). Even so, we can always ask if it is real. Or, to return to our present paper, when Hilbert introduced his consistency problem,

*We have not checked whether Columbus' hypothesis was not only false, but even absurd (when he reached the West Indies) in the sense of being easily refuted from (all) the evidence available to him about (i) the distance he himself had travelled, (ii) the length of the known sea route to India and (iii) known estimates at his time for the diameter of the earth.

what _could_ we have asked about proofs or principles of proof? Surely, no specific structural questions, say, about the _genus_ of formal derivations in systems of natural deduction! But questions about validity, in particular, consistency of principles presuppose little knowledge (of the principles one is talking about). --We certainly do not assume that questions which strike us when we know little, _never_ continue to be rewarding when we know a lot. But, conversely, we see no reason why this should happen often. (As was said already, we also cannot accept that they lose their meaning, when they have merely lost their interest.)

The second aspect of the philosophical tradition which seems to us to require attention, is the much-touted _dramatic_ claim about far-reaching consequences of our 'basic' philosophical conceptions (the latter being, allegedly, particularly 'dangerous' if they are _not_ analyzed). The matter is relevant to our own subject in several ways. In fact it seems plausible that the claim above is behind the lax standards of rigour criticized in (a), at least, in this sense: granted the claim, it would indeed be unlikely that the (pedestrian) tests which we want to apply, could be of much use; the battle is lost before we even start, because (according to the claim) those 'basic' conceptions will have 'bewitched' us (or whatever word is currently in fashion)! Be that as it may, the _empirical evidence_ of the development of the sciences--and of some proof theorists, if not of proof theory--does not seem to support the claim. (So experience in our subject may be useful for examining the general merits of the claim.) We leave it to the reader to go over the various hackneyed instances usually cited in support of the claim, and also to look for various banal counter-examples. We find it more (attractive and) instructive to go here into a case which is not clear-cut and so needs some analysis; specifically the work of Herbrand (which one of us cites frequently in this connection).

The case is interesting because, at least during his very _short_ (logical) career, there were some pretty evident connections between Herbrand's philosophical conception of mathematical reasoning and--both positive and negative sides to--his specific logical interests. We describe these connections before taking up the quite separate matter of farreaching consequences (if he had lived to see the fruits of a different, model theoretic conception).

By now it is a common place that Herbrand (or Skolem) had all the formal tools needed to prove the completeness theorem but stopped short of even formulating it because (ideological) finitist doctrines led him to reject it as senseless. On the other side of the ledger, it should be said that without those doctrines, a model theoretic proof of the socalled uniformity theorem (if $\exists \vec{x} A$, for quantifier-free A, is valid so is some disjunction $A_1 \vee \cdots \vee A_n$ where A_i is of the form $A[\vec{x}/\vec{t}]$ for terms t in the language of A) would be the natural stopping place, and there would be no reason for going on to Herbrand's theorem (which states: if d is a derivation of $\exists \vec{x} A$ there are terms $\vec{t} = \pi(d)$ for some primitive recursive π; with an elaborate version for more complex formulae than $\exists \vec{x} A$; the reasons for the elaboration are also related to the doctrine; cf. SPT II p. 130). It has been pointed out (by one of us) that even Herbrand's errors [DAA] may be related to the finitist doctrine on the hypothesis that he, in fact, knew the (easy) model theoretic proof. Since he believed that all finitistically meaningful statements have finitistic proofs anyway, there was no reason to check his own proof very carefully. (This hypothesis may of course reveal more about its proposer than its subject.)--Granted all this, Herbrand's conception may be said to have prevented him from discovering the more important completeness theorem (which, after all, was discovered independently, perhaps just because its interest is so clear), but to have led him to his own less important, but also more recondite theorem. This state of affairs is certainly consistent with what was said earlier about the heuristic value of philosophical conceptions at an early stage of a subject.

On the other hand we see no reason to suppose that Herbrand's conception would have had far reaching consequences; that it would have prevented him from learning from experience and from changing his views accordingly. In a sense--and as an example of the 'unreasonable effectiveness' of the test of formal rigour (of which we shall find further instances in (d) below)--the very errors in his work would have helped him see the defects of the conception; particularly, if our hypothesis concerning these errors is sound. We could well believe that some slow-witted or basically tired individual, terrified (consciously or unconsciously) of never having another idea in his life, might be reluctant to give second thoughts to his conceptions or to his convictions, from fear of seeing them fall to pieces without finding anything to replace them. But we find this quite implausible in the case of Herbrand.--In any case, when an individual is reluctant to change his conception is this, principally, a property of the individual or of the conception? Are there so many individuals of the type described that, statistically, the dramatic claims about the effects of philosophical conceptions and, more generally, of first impressions, are sound?

As so often with glib claims of powerful psychological 'influences' (which are so fashionable in this century), the dramatic claims above acquire a certain <u>air of unreality</u> when we try to apply them to ourselves. We certainly were not born with the conviction that abstract language was suitable for elementary metamathematics: we learnt this from experience. Or are we so exceptionally ethereal as to be immune to the temptations of false conceptions?

(d) To continue (and conclude) on an up-beat note we shall now formulate <u>positive</u> practical consequences of the reappraisals in (a)-(c). The 'negative' consequences consisted in some kind of debunking of problems which, on first impressions, seemed plausible and/or fundamental, but are not. So the 'positive' consequences may be expected to consist of problems which seemed marginal or mere refinements, but in fact are not. We consider just one topic (and only 2 of its many aspects): <u>Logical and mathematical reasoning</u>.

(i) <u>Some problems on cut-free axiomiatizations</u>. Most logicians tend to think of logic as fundamental, and of mathematics as some kind of appendage, given by some 'additional' axioms, like an after thought. Accordingly, they look for a (very) few fundamental systems such as (finitary or infinitary) first order predicate logic, and impredicative second or higher order logic. The rest would somehow 'look after itself'. This impression seems to be supported by the first crude test, namely the 'reduction' of arithmetic to logic. The test is crude inasmuch as a mere embedding of theorems is considered. Formal arithmetic can be embedded both in ω-logic, of II.1 (a special case of infinitary predicate logic), and also in (finitary) second order logic to which Peano's successor axioms are added (and the set of numbers is defined in the style of Dedekind).

It seems to us significant, that this crude test was not followed up by more searching tests, not even by the most obvious ones; obvious the moment we look for tests at all (provided only we are familiar with cut-elimination as many proof theorists are). Specifically, suppose the derivation <u>d</u> of formal arithmetic is translated into

d_∞ of ω-logic with normal form $|d_\infty|^*$

d_2 of second order logic with normal form $|d_2|^*$.

How are d_∞ and d_2 related? For example, if $|d_\infty| = |d'_\infty|$ do we have $|d_2| = |d'_2|$? In other words, do we really have a 'reduction' of arithmetic <u>reasoning</u> or only of arithmetic <u>results</u>?

To be precise, the notation above hides a problem! Not only are there distinct (terminating) cut elimination procedures with different normal forms $|d_\infty|$, but also the precise choice of defining properties of the successor function--by rules or axioms--affects the result. This last fact is obvious because cut free rules are complete for <u>validity</u>, not for <u>consequence</u> (since A cannot be derived from $A \wedge A$), and so two axiomatizations which generate the same set of theorems (by additional use of cut) need not do so if only cut-free rules are used.

Even when mathematical theories were formulated in a cut-free style (so as to be closed under cut), as has happened occasionally, this matter was not discussed as an <u>issue</u>. This omission is consistent with the idea mentioned above, of logic as 'fundamental'; the idea could hardly be convincing if the exact choice

---

* for the canonical normalization or cut-elimination procedures where, as we shall see in a moment, the word 'canonical' is a little deceptive.

of axiomatizations for mathematical theories were a major problem! (which is not solved by the allegedly 'fundamental' work on logic). Without suspecting here some conspiracy of silence, let us only note that the straightforward reaction to actual experience is different: the matter does deserve discussion. Schütte [Sch 2] gave what, no doubt, appeared to him a (if not the!) natural axiomatization STE of simple type theory with extensionality, and solemnly went on to show that this STE does not admit a normal form theorem, without even mentioning the possibility that a different axiomatization, also equivalent to STE, might admit the normal form result. Takahashi [Ta1] gave such an axiomatization but once again without formulating the issue. We see no reason to doubt generally the wordless wisdom of the East. But in the present instance there seems to be a case for a bit of explicit analysis in the Western style, to help us avoid the repetition of errors. When Uesu [U1], with Takahashi's acknowledged help, put down some (cut free) rules intended to axiomatize Zermelo's set theory, he made a formal error,* pointed out to him by Pohlers, a student of Schütte, and Uesu [U2] gave another axiomatization -still without a word on the issue involved.

Now we come to our first problem. Without asking for some grand general scheme for discovering axiomatizations (of a mathematical theory given by its set of theorems) which are complete for cut-free logical inference, we should like to see this done for a few specific theories; preferably, if the choice of axiomatization involved the mathematical content of the theory.

Remark. We no more expect 'general' rules here than for the discovery of appropriate axiomatizations in ordinary mathematics. The reader may compare (i) the present switch from one or two 'fundamental' logical systems to several (but not too many) specific ones to (ii) the passage from the (mathematically) quite useless formalization of 'all' of mathematics in a universal system (of say

---

*This is the example of the 'unreasonable effectiveness' of the test by formal rigour alluded to in (c). Uesu 'might have'-as one says-found the axiomatization [U2] by mere tinkering, and we might still be in awe of the w.w. of the East.

set theory) to the discovery of relatively few systems (for Bourbaki's structures-mère) which have greatly increased the intelligibility of mathematical reasoning.

(ii) Logic and combinatorial mathematics. As was noted in I.3(bi), logic has not found many applications in combinatorial and constructive mathematics. One theoretical point was mentioned, namely that in these domains of mathematics the set of valid formulae (in the usual logical language) is not recursively axiomatizable. In the language of $\mathfrak{K}$-founded (binary) trees: for the relevant classes $\mathfrak{K}$, the property of being $\mathfrak{K}$-founded is not recursively enumerable. Put differently, these domains of mathematics obey more 'logical' laws than those which are universally (logically) valid.. As far as proofs (of theorems in the logical language of first order) are concerned, we do better, in these domains, to use ad hoc methods than rules of inference (in the language of first order).--Realistically speaking, the same applies also to those domains of mathematics where we do have a complete formalization: modern mathematics uses--constantly and efficiently--set theoretic arguments to prove theorems (that can be) formulated in first order language, for such structures as p-adic or real closed fields. In fact the same applies to propositional logic! as suggested by examples used in the theory of infinitary propositional calculus, particularly by Stavi. Suppose $0 < n \leq N$, $0 < i \leq N+1$, $0 < j \leq N+1$ and $i \neq j$. Consider the propositional formula

$$(*) \qquad [\bigwedge_n \bigwedge_{i,j} \neg(p_{ni} \wedge p_{nj})] \wedge \bigwedge_i \bigvee_n p_{ni} .$$

We should not dream of refuting (*) by a propositional deduction, but instead we note that (*) implies the existence of a 1-1 mapping of $\{n : 0 < n \leq N\}$ onto $\{i : 0 < i \leq N+1\}$ and that this second order statement is false.

However, our formulations (of validity for models restricted to $\mathfrak{K}$ or in terms of $\mathfrak{K}$-founded trees) open up the possibility of applying metamathematics to those restricted domains of mathematics, as illustrated by the following example (to use material mentioned already in II.3(c) above).

For general $\mathfrak{K}$, we cannot expect that a formula

$$A_1 \vee \cdots \vee A_n \qquad \text{(for some finite } n\text{)}$$

is $\mathcal{L}$-valid where A_i is $A[\vec{x}/\vec{t}_i]$ and A is quantifier-free (for finite terms t_i), whenever $\exists \vec{x} A$ is $\mathcal{L}$-valid. But it does not seem implausible that, by examining elementary metamathematical proofs (of the uniformity theorem just mentioned), we can find some useful class $\mathcal{T}_{\mathcal{L}}$ of possibly infinite terms (in the language of A) such that, for each $\exists \vec{x} A$ which is $\mathcal{L}$-valid, some infinite disjunction

$$\bigvee_{i \in \omega} A[\vec{x}/\vec{t}_i]: t_i \in \mathcal{T}_{\mathcal{L}}$$

is $\mathcal{L}$-valid.

APPENDIX: KÖNIG'S LEMMA FOR FINITELY AND SEMIFINITELY BRANCHING TREES

This topic is related to the subject of the present paper in 2 ways. Quite specifically, König's lemma is mentioned repeatedly in Section I; in this connection, the results in subsection 2 below permit the formalization of various model theoretic proofs in a conservative extension of arithmetic (obtained by adding a suitable formulation of König's lemma to socalled elementary analysis). More generally, König's lemma illustrates very well a phenomenon which pervades all uses of abstract language (and not only ours for metamathematics) and which has repeatedly led to a simple minded confusion between (i) the combinatorial (mathematical) content of abstract principles and (ii) their proof theoretic strength which depends on the logical complexity (of the definitions) of the structures to which the principle is applied. We discuss the general phenomenon first because--in point of fact--it has led (one of us) to the specific work on König's lemma in subsection 2. However the latter can be read independently.

1. <u>The (logical) complexity of sets to which mathematical principles are applied</u>. (a) The case of induction. It is a commonplace that mathematicians rarely pay attention to logical complexity. A memorable illustration of this fact is provided by the apparent absurdity of the 'news' that Gentzen used ϵ_0-induction to prove the consistency of arithmetic the main (mathematical) principle of which is ordinary, that is, ω-induction. The fact that Gentzen applies ϵ_0-induction only to elementary (primitive recursive) predicates, while, in arithmetic, ω-induction is applied to all first order predicates (in the language of rings) cannot provide a <u>convincing</u> distinction unless one has doubts about the sense of logically compound expressions. An equivalent way of putting the matter uses a two sorted language, for numbers and predicates or sets of numbers. The <u>abstract</u> induction principle

$$\forall X\{[0 \in X \rightarrow \forall x(x \in X \rightarrow sx \in X)] \rightarrow \forall x(x \in X)\}$$

is supplemented by independent set existence axioms expressing closure under say primitive recursion and the 'logical' operations (that is, <u>projection</u> since union and complementation happen to be primitive recursive). The system corresponding to Gentzen's use of ϵ_0-induction supplements the abstract principle (of ϵ_0-induction) only by the non-logical set existence axioms. But why should one 'leave out' the logical ones?

A good reason would seem to be provided by (manageable) models which satisfy the non-logical axioms but are not closed under projection--and the collection of (primitive) recursive sets is certainly such a model (when ω is the domain of the individual variables and $\underline{s}$ the successor). But the 'reason' is deceptive because this model is an ω-model and so automatically satisfies induction without restriction on the X.* It must be admitted that we do not know, at present, any models that are not standard w.r.t. the integers (that is, not ω-models), satisfy induction for predicates of limited complexity, and are sufficiently transparent to be of intrinsic interest. In other words, we do not have here a clear parallel to the typical situation of ordinary axiomatic mathematics where the study of restrictions on the 'usual' axioms (for example, in the passage from Dedekind's second order axioms for $\mathbb{R}$ to the theory of real closed fields) provides obviously interesting structures which satisfy the restricted but not the usual axioms.

As already mentioned in II, we regard the subject of $\mathfrak{L}$-founded derivations in the present paper as providing a 'home' for Gentzen's mathematical ideas. Here the restiction on the logical complexity of the predicates to which (bar) induction on $\mathfrak{L}$-founded trees can be applied, is certainly not arbitrary! The principle is simply not valid unless, roughly speaking, the predicates $\in \mathfrak{L}$. So all that remains is to make clear the interest of the $\mathfrak{L}$-founded derivations themselves, which was done in I.3(a) without any logical sophistry.

To avoid misunderstanding it is perhaps worth making two remarks. First of all, even though we do not have any really manageable non-standard models,

---

*This is by no means an isolated phenomenon. For example, β-models automatically satisfy <u>bar induction</u> without any restriction on the predicates to which this principle is applied; only the (partial) ordering along which the induction proceeds is required to be a set in the domain of the β-model.

there are a few very elegant uses of (the language of) such models anu some superficial properties (common to all such models), for example, the indefinability of the set of standard integers of the model: we know this much, without knowing anything about its genuinely mathematical properties. One such use, due to Scott [Sc], is described in subsection 2: it is particularly appropriate because its principal purpose is to define a good ω-model (for elementary analysis to which König's lemma is added).

The second remark concerns the study of restrictions on the predicates to which abstract principles such as induction are applied. As already mentioned there is a logical view which requires the restrictions, for example because of (genuine or ethereal) doubts about the existence of sets having certain formal properties. But this view is not at all necessary to make the study in question sensible! Whether we like it or not, any given application of an abstract principle can only involve a finite number of predicates. Moreover, in practice the predicates will be limited by the mathematicians' view of the problem which he is trying to solve by means of the abstract principle. If he is thinking number theoretically, the predicates will be expressed in the language of number theory; if he is thinking in a broader context, for example of the integers embedded in the complex planc, the predicates will be in the language of function theory, and the additional set existence axioms will be those familiar in the given context. It certainly can do no harm to have some idea of the consequences of a given 'view', for example, for the class of, say, number theoretic theorems provable by means of the abstract principle. This so to speak negative interest of non-standard models seems to us quite real.*

(b) Trees and branchings. To fix ideas let us consider trees T whose nodes which are all at a finite distance from the root, are labelled by integers. König's lemma, in ordinary mathematical language, asserts this:

---

*If this is accepted, one can at the same time (i) accept the intuitive impression that non-standard models are of interest and (ii) dismiss the contention that non-standard models discredit our usual conception of the (standard) integers.

If T is unbounded and each node of T is either terminal or has a finite number of immediate descendants (= predecessors in the tree ordering) then T has an infinite path.

For example, if T is binary, that is, each node has at most 2 predecessors, and labelled 0 or 1, then any infinite path π can be coded by a set $X_\pi \subset \omega$ as follows:

$$n \in X_\pi \iff \text{the label at the nth node of } \pi \text{ is } 1$$

$$n \notin X_\pi \iff \text{the label at the nth node of } \pi \text{ is } 0.$$

To formulate the principle in set theoretic language, specifically in the usual language of sets of natural numbers, we shall regard T as a predicate of finite sequences of 0, 1, closed under initial segments. We then have a choice between

(i) the abstract principles together with appropriate existential set theoretic axioms or

(ii) a schema where T is defined by some given class of formulae in the language, in the style of the comprehension schema.

Evidently (ii) is sensitive to the class of formulae chosen; for an arbitrary predicate P, we can obviously arrange T to define a tree consisting of a single path π s.t. $X_\pi \leftrightarrow P$; in short, the comprehension principle follows from the schema (ii) (by use of quite elementary axioms).

As to (i) it has long been known that the abstract principle is satisfied by arithmetically definable sets in the sense that, if $T \in \Pi^0_\infty$ so does some X_π; and the first basis theorem in the literature asserts that if T is recursive some $X_\pi \in \Delta^0_2$. Also it has long been known that if T is recursive then there need be no recursive X_π.

Both the positive and negative results apply equally to the two types of (recursive) trees mentioned above, namely those labelled by 0 or 1 and those labelled by integers without any bounds on the labelling. However, there is an obvious difference between these two types. If T is recursive and the labels are 0 or 1, the decision whether a given node is terminal, has one or two descendants is also recursive. If, in contrast, the labels are not bounded and even if we are given that the tree is binary (in the sense that each node has at most 2 descendants) there is no obvious reason why the decision above should be recursive: the relation between an (arbitrary) node and its immediate descendants is, at least prima facie, only r.e. Put differently, we have no reason to expect recursive isomorphisms between the two classes of recursive binary trees which have bounded, resp. unbounded labels. Some years ago one of us presented this state of affairs to H. Friedman who gave a very satisfactory answer. The abstract version of König's lemma for binary trees with unbounded labels together with closure under a few primitive recursive operations is equivalent to the full arithmetic comprehension principle (with parameters), and thus not a conservative extension of first order arithmetic.--For reference in (d) of subsection 2 below, it is worth noting that though Friedman's result is very satisfactory for a good choice between bounded and unbounded labels on finitely branching trees, it also casts doubt on the mathematical interest of such equivalence results (w.r.t. to elementary analysis): it is quite clear that qua combinatorial principle, König's lemma has nothing to do with the arithmetic (or any other) comprehension principle.

Exercise. The reader may wish to derive formally, from König's lemma for unboundedly labelled finitely branching trees, that

for any ordering R, R is well-founded if and only if 2^R is well-founded.

where 2^R has as domain finite descending sequences (in order R) ordered lexicographically (w.r.t. R). This observation, which we learnt from W. A. Howard, also shows that König's lemma for unboundedly labelled finitely branching trees (together with elementary analysis) is not conservative over first order arithmetic. Here it seems quite clear--and, perhaps, worth making precise--that, qua combinatorial principle, König's lemma does have something to do with the step from R to 2^R above.

In contrast to Friedman's result, as mentioned in subsection 2 the principle for bounded labels etc. is conservative over arithmetic. Naturally, the restriction to the particular labels 0, 1 is immaterial: it is sufficient to have a recursive bound, say $\alpha(n)$, for the labels at the nodes at distance n from the root of the tree; since our tree is recursive such a bound allows the recursive determination of all the labels.

Remark on an alternative analysis by Troelstra [Tr], for systems in the language of natural numbers and functions, but using intuitionistic rules of logic. Leaving aside some technical refinements, the most striking difference is this.

> At least formally, the intuitionistic version* is insensitive to the logical complexity of the definition of T, and to the distinction between the two types of labelling. (We shall call the case of bounded labels: finite branching, the other: semi-finite branching.)

Moreover--and this is of course the reason for going into the matter at some length--this phenomenon is by no means isolated. For example, in classical logic the proof theoretic strength of choice schemata depends sensitively on the logical complexity of the relation, say R, in

$$\forall x \exists y R \rightarrow \exists f \forall x R[y/f(x)] ,$$

or in bar induction etc.; in intuitionistic logic of functions (not of species!) generally speaking the opposite is true. The intended intuitionistic interpretation treats functions, including the (logical) operations implicit in the existential quantifier or in disjunction, in a quite narrow sense, and asymmetrically from species (which need not have characteristic functions--and the usual formal systems for intuitionistic mathematics permit an interpretation in which the functions are recursive**). Consequently, even if the tree T is

* of: if every path through T is finite then T itself is finite, where a path is given by a function (of n) naming the nth label on the path.

** at least as long as the logical operations are reinterpreted in the sense of realizability.

defined by a logically complicated expression, this is so to speak nullified by the requirement of finite branching (in the sense that, for each node, there is either no descendent or one or two).--The reader may wish to verify, that, for logically complex T, a mere bound on the number of descendents would alter the situation completely.

2. ω-models of König's lemma (for finitely branching trees) derived from non-standard models of first order arithmetic. The models considered below were introduced by Scott [Sc], also in connection with König's lemma (for binary trees with labels 0 or 1). Naturally, 15 years later we pay attention to aspects of these models, and consider variants, not treated in the original publication. We begin with a description of the models, then look at some properties or so-called 'axioms' which they do or do not satisfy, and finally make some relevant applications.

(a) We start with any complete and consistent extension of first order arithmetic P; in short, a model of P in the sense of Section I. 1 . As is well-known, there are such models which are arithmetically definable, even in Δ_2^0, in such a way that the nth natural number of the model, say i(n), is given by a primitive recursive function of n. Furthermore, by the formalized version of the completeness theorem (adapted to Henkin's proof, from the presentation in Hilbert-Bernays for Gödel's proof which gives a Δ_2^0 satisfaction predicate only on the atomic formulae, not on all formulae), the Δ_2^0 satisfaction relation so obtained can be formally proved in P to be a model of P from the assumption that P is consistent, Con(P) for short. Naturally since the property of being a model of P is a single sentence, the proof is given in a subsystem of P. (The reader interested in such matters should note that 'being a model of P' is expressed by infinitely many sentences if a model is given by its satisfaction relation for atomic formulae.)

The class of sets (of natural numbers) or relations (between natural numbers) to be considered are what Scott called 'binumerable' in the given extension; we prefer to call them more simply: the sets and relations definable (in the given model of P) <u>on</u> ω,

that is, explicitly definable by a formula of P, say F, 'without parameters': $\{n : i(n) \text{ satisfies } F\}$.

As is well-known all recursive sets are definable <u>on</u> ω in all models of P (and, indeed, all those sets are definable in all models of quite small subsystems of P, for example, Q). Not all r.e. sets are necessarily definable <u>on</u> ω in each such model since there are models of degree < 0'.

The class of sets considered contains not only all recursive sets but is <u>closed</u> under recursive operations; more formally, it satisfies

$$\Delta_1^0 - CA : \forall RS[\forall n[\exists pR(n,p) \longleftrightarrow \neg \exists qS(n,q)] \longrightarrow \exists X \forall n[X(n) \longleftrightarrow \exists pR(n,p)]]$$

where R, S are variables over binary relations and X is a variable over sets.

To see this, suppose R and S are defined <u>on</u> ω, by the formulae F_R and F_S with the free variables x, y; resp. x, z. Then X is defined <u>on</u> ω by

$$\exists y[F_R \wedge (\forall z \leq y) \neg F_S]$$

Here again one does not use the full 'force' of P, but just enough axioms to ensure (by Rosser's condition) that all the models are <u>end extensions</u> of ω.

<u>Exercise</u>. The reader should write down variants of König's lemma (for finitely branching trees) which are satisfied in the model, and consider their logical relations; any such tree being given by a set of (formally finite) paths closed for initial segments together with a bounding function for the labels at level <u>n</u>. In particular, consider the variants obtained by taken for the trees (i) sets and functions definable on ω in the complete extension of P considered, (ii) sets and functions which include all those that are recursive in parameters which are definable outright.

(b) Defining trees (and a reminder of the definition of functions <u>on</u> ω in contrast to: definitions of their graphs). Both naively and for the specific purpose of satisfying König's lemma, it is natural to use definitions, say T, of trees $\mathcal{T}$ which define, in our (nonstandard) model of P, <u>end extensions</u> of the restriction of $\mathcal{T}$ to ω, that is

> any initial path of $\mathcal{T}$ of length <u>n</u> where <u>n</u> is a natural number has labels which are also natural (= not: nonstandard) numbers.

(Here, as in subsection 1 above, a tree is given by its formally finite initial paths.) Clearly, a standard tree is recursive in the sense that the set of finite initial sequences (coded canonically) is recursive and if there is a recursive bound for the labels at level $n \in \omega$, there is a formula T which defines, in each (non-standard) model of P, an end extension of the given tree. In particular, this is clear for trees whith labels 0 or 1 only.

<u>Exercise</u> (for a reader fond of 'omitting type theorems'). Show that if a tree $\mathcal{T}$ (is recursive and, say, binary but) does not have a recursive bound for the labels at levels $n \in \omega$, there is a model of P in which no formula defines an end extension of $\mathcal{T}$.

Thus our requirement on finitely and not merely semi-finitely branching trees makes good sense in terms of (definitions applied in) non-standard models; even if we do not mind what the definition 'does' to sequences of non-standard length, we do not want non-standard labels at a standard, genuinely finite distance from the root of the tree. (The 'extension' of definitions T to non-standard models illustrates modern mathematical practice discussed in I.3(c).)

The 'reminder' referred to above, concerning <u>functions</u> f <u>on</u> ω, involves <u>uniqueness</u> properties of definitions in the whole model, not only on ω. The <u>graph</u>, say G_f of f, is said to be defined <u>on</u> ω by F provided only $(n,m) \in G_f \Longleftrightarrow \langle i(n), i(m) \rangle$ satisfies F; evidently, since G_f is functional <u>on</u> ω, for each $\underline{n} \in \omega$, only $m = f(n)$ will do, but the e may be non-standard <u>x</u> s.t.

$$\langle i(n), x \rangle \text{ satisfies } F.$$

A <u>function</u> f is said to be defined by F if, in addition, for each $n \in \omega$

$$\forall x\{[\langle i(n), x\rangle \text{ satisfies } F] \Longleftrightarrow x = i(fn)\}$$

is true in the model.

(c) König's lemma holds in the models considered in (a), for trees $\mathfrak{T}$ defined by formulae T described in (b).

Suppose $\mathfrak{T}$ is unbounded (in ω). Since P contains induction, in any non-standard model (of P), T must be satisfied by some path π of non-standard length. Thus $\mathfrak{T}$ has an infinite path definable on ω by a formula _with_ parameters (in the model), since the proper segments π of standard length are themselves standard and therefore in $\mathfrak{T}$, and so the restriction of π to ω goes through $\mathfrak{T}$. Here we have used our choice of definitions in (b) for finitely branching trees which excludes non-standard labels at a finite distance from the root of $\mathfrak{T}$.

To get a definition _without_ parameters one distinguishes between two cases:

(i) The (nonstandard) tree defined by T has a, necessarily, nonstandard bound, that is, it is finite in the sense of the nonstandard model. Then, since induction holds, there is a highest level, and we take the node, say N, with the smallest label at that highest level. Having just given an explicit definition (without parameters), we also have a definition of the path joining N to the roots of the tree.

(ii) The tree is infinite in the sense of the (nonstandard) model (and, formally, finitely branching). Now we take the leftmost _infinite_ branch of the tree defined by T.

So in each of the cases there exists a formula which defines, _on_ ω, an infinite path through $\mathfrak{T}$. (The reader who does not like undecided disjunctions, will easily find a _single_ formula to define a path; let F_T express formally that the tree defined by T is finite, and let $P^{(i)}$ and $P^{(ii)}$ be the explicit definitions of paths given in (i) and (ii) resp. Then take

$$(F_T \wedge P^{(i)}) \vee (P^{(ii)} \wedge \neg F_T) .)$$

Actually, for the particular applications mentioned below it would be sufficient to consider the class of sets definable by use of (nonstandard) parameters, or to consider models in which each element is definable outright .

(d) Applications and references to the literature. An obvious application, with a comic side to it, is this. Let Δ_1^0-CA be the subsystem of analysis (also called second order arithmetic) in which induction is applied to arbitrary formulae, but comprehension is restricted to Δ_1^0-formulae as in (a), and let KL be König's lemma for finitely branching trees in abstract form, as described in 1(a) above. Then we have a proof in primitive recursive arithmetic of

$$\mathrm{Con}(P) \longrightarrow \mathrm{CON}(KL + \Delta_1^0 - CA) .$$

Since $KL + \Delta_1^0 - CA$ literally contains P, this means that Con(P) cannot be proved in $KL + \Delta_1^0 - CA$.

As one of us reported in a review of [G], Gentzen's original version of his consistency proof for (P) was criticized for having used, allegedly, KL. The criticism overlooked (among other things) the crucial point, discussed in 1, that the strength of KL is very sensitive to the complexity of the trees to which the abstract principle is applied; for example by 1(b) one gets a trivial proof of Con(P) by applying KL to e.g. arithmetic trees, and by Friedman's observation mentioned there, also by applying KL to recursive but semifinitely branching trees. By the remark above, Con(P) cannot be proved by use of KL applied to finitely branching recursive trees.

More amusingly,* when Spector applied the abstract principle of bar induction (for infinitely branching trees) to trees labelled by objects of higher type, first reactions essentially repeated the oversights involved in the earlier criticism of Gentzen's work. It was 'argued' that it can make no difference whether the principle is applied to trees of natural numbers or to trees of objects of higher type

* at least according to Marx's sense of humour applied to patterns in history: the first instance (of a pattern) may be tragic, any repetition is comic. cf. also Walpole whose man of feeling apparently does not recognize a repetition as such.

(where, incidentally, it was far from clear just which operations of higher type were to be meant; especially since, demonstrably, not all such operations were admissible; for example, the operations had to be extensional and even continuous --for the product topology at types $(0 \rightarrow \sigma) \rightarrow 0$). -The reader may also wish to pursue the parallel in subsection 2 where invariance properties of definitions of trees (at finite levels) were found to be important: in the case of trees of higher type objects, the same formula will define different trees according to the class of such objects present in the model considered (even if only ω-models are allowed).

However inappropriate these first reactions may have been in the logical contexts (of consistency proofs for arithmetic and analysis respectively) in which they occurred, they are obviously symptoms of the conflict between mathematical interest and (traditional) logical analysis which is part of the theme of the present paper. After all, if naively we wish to convey the idea of a proof, we do so by mentioning the abstract combinatorial principle used and, perhaps, the 'kind' of predicate to which it is applied; it is only the particular classification, by means of logical complexity, that is rarely convincing. (And, if, occasionally, it is, for example, when Σ_1 or Δ_1 is connected with recursion theory or invariance, the discovery of the particular connection constitutes the principal interest; not merely the fact that Σ_1 is a stage in the hierarchy Σ_n for $n = 1, 2, \ldots$*).

As a practical consequence, for further research in the present area, we are led to be more--or, perhaps, too-discriminating w.r.t refinements which we consider to be rewarding. As an example, we consider the improvement of the relative consistency results (of $KL + \Delta_1^0 - CA$ over P) to:

$$KL + \Delta_1^0 - CA \text{ is conservative over } P,$$

also established by primitive recursive (metamathematical) methods; this is an improvement since relative consistency asserts that $KL + \Delta_1^0 - CA$ is conservative

*Put differently, we do not think of the 'simple minded' step from Σ_1 to Σ_n as a particularly significant direction for generalizations.

over P w.r.t. the single formula 0 = 1. There is a considerable literature on this subject, reviewed in [Tr]. Also the work in (a)-(c) above can be modified to yield the conservation result by considering, in place of P, finite subsystems P_n and models of P_n established in P itself. (One has to keep count of the complexity of the formulae in (c) to which induction is applied.)

While, from the point of view of logical complexity, P is as natural a system as one can reasonably expect (in view of the incompleteness theorems), we reamin skeptical:

What does P have to do with KL?

One has the impression that certain 'optimal' subsystems P_0 of P are much more relevant; not because they are 'weaker' or 'smaller' but simply because the proof of the result for P itself abounds with formulae in the 'neighbourhood' of Δ_2^0. So we look for a statement (of another result!) which is both memorable and sums up this feature of the analysis. Roughly speaking, we expect to find 'subsystems' P', not necessarily in the language of P itself, such that the 'corresponding' systems EL' of elementary analysis together with KL are conservative over P'. We should not be surprised to find that some of the 'subtler' hierarchies of Δ_2^0 could be used to formulate sharp results here.

As always when 'subtle' or pedantic distinctions appear to be necessary for a proper statement of the facts, it may be time to look at bigger things which magnify the distinctions (and thus remove the horrors of pedantry). In the particular case of KL, we should pass from the language of number theory to that of set theory, or better still (since we have found $\mathcal{L}$-founded derivations useful) to the language of trees which are--unlike sets--not well-founded by definition. This last passage is also suggested by work on infinitary languages $\mathcal{L}_A$ for admissible sets A: the analysis of the roles of admissibility, for example by Barwise and Stavi, uses closure conditions on A which are extremely natural in the language of sets; but we hardly ever use the corresponding tree-theoretic operations.

REFERENCES

[A] Ackermann, W., Die Widerspruchsfreiheit der allgemeinen Mengenlehre, Math. Ann. 114 (1937) 305-315.

[C] Carstengerdes, W., Mehrsortige logische Systeme mit unendlichlangen Formeln, Archiv Math. Logik Grundlagenforsch 14 (1971) 38-53 and 108-126.

[CK] Chang, C.C. and Keisler, J.H., Continuous model theory, Princeton, 1966.

[DAA] Dreben, B., Andrews, P., and Aandera, S., False lemmas in Herbrand, Bull. AMS 69 (1963) 699-706.

[F] Feferman, S., Lectures on proof theory, Springer Lecture Notes 70 (1968) 1-108.

[Fr] Friedman, H., Iterated inductive definitions and Σ^1_2-AC, pp. 435-442 of: Intuitionism and Proof Theory (ed. Myhill et al.) North Holland Publ. Co., 1970.

[G] Gentzen, G., The collected papers of Gerhard Gentzen, ed. M.E. Szabo, Amsterdam 1969; rev. J. of Philosophy 68 (1971) 238-265.

[Gi] Girard, J.-Y., Three-valued logic and cut-elimination: the actual meaning of Takeuti's conjecture, Fund. Math. (to appear).

[J-S] Jockusch, C.G. and Soare, R.I., Π^0_1-Classes and degrees of theories. Trans. A.M.S. 173 (1972) 33-56; rev. Zbl. 262 (1974) 19; no. 02041.

[KK] Kreisel, G. and Krivine, J.-L., Elements of mathematical logic, second revised printing, North Holland Publ. Co., 1971.

[KL] Kreisel, G. and Levy, A., Reflection principles and their use for establishing the complexity of axiomatic systems, Zeitschrift math. Logik Grundlagen 14 (1968) 97-142.

[LE] Lopez-Escobar, E.G.K., On an extremely restricted ω-rule, Fund. Math. (to appear).

[Mi] Mints, G.E., On E-theorems, Zapiski 40 (1974) 110-118.

[ML] MartinLöf, P., Hauptsatz for the intuitionistic theory of iterated inductive definitions, pp. 179-216 of Proc. Second Scand. Logic Symposium (ed. Fenstad) North Holland Publ. Co., 1971.

[Mo] Mostowski, A., On recursive models of formalized arithmetic, Bull. Acad. Pol. Sc., cl. III, 5 (1957) 705-710.

[OL] Kreisel, G., Ordinal logics and the characterization of informal concepts of proof, pp. 289-290 in Proc. ICM Edinburgh 1968.

[P] Parsons, C., Transfinite induction in subsystems of number theory (abstract), JSL 38 (1973) 544.

[Sch] Schütte, K., Syntactical and semantical properties of simple type theory, JSL 25 (1960) 305-326.

[Sch 1] ____, Beweistheorie, Berlin, 1960.

[Sch 2] _____, On simple type theory with extensionality, pp. 179-184 in: Logic, Methodology and Philosophy of Science III, North Holland Publ. Co., 1968.

[Sco] Scott, D.S., Algebras of sets binumerable in complete extensions of arithmetic, pp. 117-121 of: Proc. Symp. Pure Math. 5, AMS, 1962.

[Sh] Shoenfield, J.R., Mathematical logic, Addison-Wesley, 1967.

[SPT] Kreisel, G., A survey of proof theory, JSL 33 (1968) 321-388.

[SPT II] _____, A survey of proof theory II, pp. 109-170 of: Proc. Second Scand. Logic Symp. (ed. Fenstad), North Holland Publ. Co., 1971.

[St] Statman, R., Structural Complexity of proofs, Dissertation, Stanford, 1974.

[Ta 1] Takahashi, M., Simple type theory of Gentzen style with the inference of extensionality, Proc. Jap. Acad. 44 (1968) 43-45.

[Ta 2] _____, Many valued logics of extended Gentzen style II, JSL 35 (1970) 493-528.

[Tr] Troelstra, A.S., Note on the fan theorem, JSL (to appear).

[U 1] Uesu, T., Zermelo's set theory and G*LC, Comment. Math. Univ. St. Pauli 16 (1967) 69-88.

[U 2] _____, Correction to 'Zermelo's set theory and G*LC', ibid. 19 (1970) 47-49.

[V] Vaught, R.L., Sentences true in all constructive models, JSL 25 (1960) 39-53.

PS (suggested by last-minute correspondence between two of the authors) concerning mainly the roles of <u>continuity</u> and <u>$\mathcal{L}$-foundedness</u> for analyzing cut-elimination procedures, especially in Part II. In §4 of the Introduction (pp. 8-9) and elsewhere (e.g.p. 51), continuity is considered as an <u>additional</u> requirement (p. 9, l. 5) to be satisfied by 'standard procedures' (p. 51, l.-4); the ultimate aim being to use the refined cut elimination procedures for efficient solutions (p. 9, l. 12) of the kind of open problems stated near the end of II.2(d) (cf. p.54, l.-3). It is quite clear that these problems are not solved by what were called 'trivial' procedures (in the first paragraph of §4 on p. 8), namely procedures obtained automatically from normal form theorems(or, equivalently, the completeness of the cut free rules for the class of models considered). Through an oversight we failed to discuss explicitly the obvious pedagogic question whether <u>continuity and/or $\mathcal{L}$-foundedness considerations alone are sufficient to distinguish between trivial and standard procedures</u>. As will be seen in (1) and (2) below, the answer is <u>negative</u>.

Correction (of a second pedagogically serious omission at the beginning of §4 on p. 8). There are two familiar trivial procedures. The first, described loc. cit., involves running through all cut-free derivations; this type of procedure is patently discontinuous in the case of infinite derivations. The second trivial procedure involves simply the canonical 'refutation' trees (for a formula A) called T_A^{CF} in the finite case (§2 on p. 22) and $T_A^{\omega CF}$ in the infinite case (p. 39); given any derivation tree, T', with the end formula $E(T')$ one has the procedures:

$\rho_2 : T' \mapsto T_{E(T')}^{CF}$ in the finite case and $\rho_2^{\omega} : T' \mapsto T_{E(T')}^{\omega CF}$ in the infinite case.

We failed to mention the second procedure where it should have been stressed (§4 on p. 8). It is plausible that, at least statistically, ρ_2 is quite efficient in the finite case in this sense: for constructing a cut free derivation of A, it may be statistically efficient to make a fresh start, to construct T_A^{CF} once one knows that A is valid, instead of trying to convert a given derivation with cut into one without cut by means of a 'standard' procedure. We now consider ρ_2^{ω}

1. Evidently, ρ_2^{ω} is continuous.

2. Also (as will be shown elsewhere) ρ_2^{ω} is essentially optimal as far as $\mathcal{L}$-foundedness is concerned. Roughly speaking, for any A, $T_A^{\omega CF}$ is optimal among all cut free derivations of A in the sense that, for $\mathcal{L}$ satisfying a few closure conditions, $T_A^{\omega CF}$ will be $\mathcal{L}$-founded provided there is any cut free $\mathcal{L}$-founded derivation of A at all.

In short, the Pedagogic Remark (bottom of p. 54) which refers to differences between the authors' views, understates the extent to which the value of our analysis depends objectively on the applications alluded to there. For if continuity and $\mathcal{L}$-foundedness were the only conditions imposed on cut-elimination procedures, the trivial 'procedure' ρ_2^{ω} would do, despite the fact that it is completely independent of the structure of the derivation trees to which it is applied (except for the end formula).

Concerning applications, it is evident that ρ_2^{ω} is inadequate for anything like E - theorems of [Mi]. (Here it should perhaps be mentioned that the last 8 lines of p. 64 mention only a small fraction of what is proved in [Mi], since the same t_d is also given by various functional interpretations of $\underline{d}$, and so the interpretations of $A[t_d]$ are derivable too. Clearly ρ_2^ω is inadequate even w.r.t. this 'small fraction'). In addition it seems plausible that ρ_2^ω is also inadequate for ensuring the separation of ordinal assignments from the cut elimination procedure itself, in other words, for establishing that derivation trees need not be 'enriched' by such assignments (for a structural theory of proofs); ef. bottom of p. 14 and p. 60(c). The matter is, at least prima facie, of obvious pedagogic interest since, by p. 87, Gentzen introduced the whole business of ordinal notations as a result of a misunderstanding and so a certain skepticism concerning the significance of such notations is justified. At the same time the separation would support the impression that Gentzen's work in toto is significant, only the roles of (i) the well-known business of ordinal notations and (ii) of the almost anonymous details of the structural transformations themselves would be reversed. (Projected back to the finite case, ordinal assignments involve nothing more than the length of the longest paths of the derivations considered).

Some remarks (to avoid possible misunderstandings). (a) Mathematically, we do not wish to claim that ours is the only reasonable style of analyzing differences between trivial and standard cut elimination procedures; for the reasons given at the bottom of p. 33 and expecially on top of p. 34, some of us find our use of trees that are not necessarily well founded promising for getting efficient procedures for operating on large finite trees. At least for one of us this constitutes a change of an earlier view (in [SPT]) stressing generalizations to uncountable formulae, but with well founded proof trees. As J. Stavi has pointed out to us, here (the obvious generalization of) ρ_2 cannot be used, simply because T_A^{CF} would not be determined for uncountable A (nor for an A for which there is no enumeration of its subformulae in the, say, admissible set from which the syntactic objects of the language are taken). (b) Bibliographically, we do not claim that pp. 55-57 give an even remotely adequate summary of the Moscow Symposium paper loosely discussed in II.2(d) (or, more precisely, of the detailed analysis by one of us of a specific formal system, which is behind that paper). It may be superfluous to add that, apart from the author who actually wrote II.2(d) or, for that matter, Section III, we have stronger reservations about his critical discussion than about the paper under discussion (which, after all, establishes precisely what it says it does).

PPS. Though, of course, the list of possibly significant differences between 'standard' cut-elimination procedures (p.51) and ρ_2 or ρ_2^{ω} (p. 91a) is almost endless, the following should perhaps be stressed; particularly in view of the basic problem on p. 38 (and as a counterpart to (2) on p. 91a). Since the canonical trees T_A^{CF} and $T_A^{\omega CF}$ code all countable counter models to A up to isomorphism, in general $\rho_2 T'$, resp. $\rho_2^{\omega} T'$ will code counter models of $E(T')$ not coded by T'. (Since we consider $\mathcal{L}$-founded, not necessarily well-founded derivation trees T', $E(T')$ may have some counter models). We should expect some procedure ρ 'like' the standard procedures to solve the problem on p. 38 positively.

WEAK MONADIC SECOND ORDER THEORY OF SUCCESOR IS NOT ELEMENTARY-RECURSIVE[†]

Albert R. Meyer

Let L_{SIS} be the set of formulas expressible in a weak monadic second order logic using only the predicates $[x = y+1]$ and $[x \in X]$. Büchi and Elgot[3,4] have shown that the truth of sentences in L_{SIS} (under the standard interpretation $< N$, successor $>$ with second order variables interpreted as ranging over finite sets) is decidable. We refer to the true sentences in L_{SIS} as <u>WSIS</u>. We shall prove that <u>WSIS</u> is not elementary-recursive in the sense of Kalmar. In fact, we claim a stronger result:

<u>Theorem 1</u>: There is a constant $\epsilon > 0$ such that if $\mathfrak{M}$ is a Turing machine which, started with any sentence in L_{SIS} on its tape, eventually halts in a designated accepting state if and only if the sentence is true, then for all sufficiently large n, there is a sentence of length n[†•] for which $\mathfrak{M}$'s computation requires

$$\left.2^{2^{\cdot^{\cdot^{\cdot^{2^n}}}}}\right\}\lfloor \epsilon \cdot \log_2 n \rfloor$$

steps and tape squares.

[†•] By the length of a sentence we mean the number of characters in it including parentheses, digits in subscripts, etc. Any of the standard conventions for punctuating well-formed formulas may be used, except that in some cases conventions for matching parentheses may imply that for infinitely many n, there cannot be any wff's of length n. In this case, we assume that wff's may be lengthened by concatenating a finite sequence of "blank" symbols which leave the meaning of the wff unchanged, so that sentences of length n can be constructed for all sufficiently large n.

[†] Work reported here was supported in part by Project MAC, an M.I.T. research program sponsored by the Advanced Research Projects Agency, Department of Defense, under Office of Naval Research Contract Number N00014-70-A-0362-0006 and the National Science Foundation under contract number GJ34671 Reproduction in whole or in part is permitted for any purpose of the United States Government.

Let $t_0(n) = n$, $t_{k+1}(n) = 2^{t_k(n)}$. A well-known characterization of the elementary-recursive functions by R.W. Ritchie[14] shows that a set of sentences is elementary-recursive iff it is recognizable in space bounded by

$$t_k(n) = \left. 2^{2^{\cdot^{\cdot^{\cdot^{2^n}}}}} \right\} k$$

for some fixed k and all inputs of length $n \geq 0$. Hence, WSIS is not elementary-recursive.

In these notes we prove a somewhat less powerful version of Theorem 1, which by Ritchie's result is still sufficient to establish the truth of our title.

Theorem 2: Let $\mathfrak{M}$ be a Turing machine which, started with any sentence in L_{SIS} on its tape, eventually halts in a designated accepting state iff the sentence is true. Then for any $k \geq 0$, there are infinitely many n for which $\mathfrak{M}$'s computation requires

$$\left. 2^{2^{\cdot^{\cdot^{\cdot^{2^n}}}}} \right\} k$$

steps and tape squares for some sentence of length n.

The idea behind our proof will be to show that there are sentences in L_{SIS} of length n which describe the computation of Turing machines, provided the space required by the computation is not greater than $t_k(n)$. Since a Turing machine using a given amount of space can simulate and differ from all machines using less space, we will deduce that small sentences in L_{SIS} can describe inherently long computations, and hence L_{SIS} must itself be difficult to decide.

Actually it will be more convenient to develop an intermediate notation called γ-expressions for sets of finite sequences. We will show that γ-expressions can, in an appropriate sense, describe Turing machine computations, and that L_{SIS} can describe properties of γ-expressions.

Definition: Let Σ be a finite set whose elements are called symbols. Σ^* is the set of all finite sequences of symbols from Σ. For $x, y \in \Sigma^*$, the concatenation of x and y, written $x \cdot y$ or xy, is the sequence consisting of the symbols of x followed by those of y. An element $x \in \Sigma^*$ is called a word, and the length of x is written $\ell(x)$. We use λ to designate the vacuous sequence of length zero in Σ^* which by convention has the property that $x \cdot \lambda = \lambda \cdot x = x$ for any $x \in \Sigma^*$. (Σ^* is the free monoid with identity λ generated by Σ.) Concatenation is extended to subsets $A, B \subset \Sigma^*$ by the rule

$$A \cdot B = AB = \{xy \mid x \in A, y \in B\}.$$

For any $A \subset \Sigma^*$, we define

$$A^0 = \{\lambda\}, \ A^{n+1} = A^n \cdot A, \ A^* = \bigcup_{n=0}^{\infty} A^n.$$

These operations are familiar in automata theory. We introduce one further mapping.

Definition:[†] For any Σ, the function $\gamma_\Sigma \colon P(\Sigma^*) \to P(\Sigma^*)$ is defined by the rules

---

[†] $P(S) = \{A \mid A \subset S\}$ = the power set of S.

$$\gamma_\Sigma(\{x\}) = \{y \in \Sigma^* \mid \ell(x) = \ell(x)\} = \Sigma^{\ell(x)} \qquad \text{for } x \in \Sigma^*,$$

$$\gamma_\Sigma(A) = \bigcup_{x \in A} \gamma(\{x\}) \qquad \text{for } A \subset \Sigma^*.$$

We omit the subscript on γ_Σ when Σ is clear from context.

γ-expressions over Σ are certain words in $(\Sigma \cup \{\underset{\sim}{\gamma}, \underset{\sim}{\cdot}, \underset{\sim}{\neg}, \underset{\sim}{\cup}, \underset{\sim}{(}, \underset{\sim}{)}\})^*$ where $\underset{\sim}{\gamma}, \underset{\sim}{\cdot}, \underset{\sim}{\neg}, \underset{\sim}{\cup}, \underset{\sim}{(}, \underset{\sim}{)}$ are symbols not in Σ. Any γ-expression α defines a set $L(\alpha) \subset \Sigma^*$.

Definition: For any Σ, γ-expressions over Σ and the function $L:\{\gamma\text{-expressions over } \Sigma\} \to P(\Sigma^*)$ are defined inductively as follows:

1) σ is a γ-expression over Σ for any $\sigma \in \Sigma$, and $L(\sigma) = \{\sigma\}$;
2) if α, β are γ-expressions over Σ, then $\underset{\sim}{(}\,\alpha \underset{\sim}{\cdot} \beta\,\underset{\sim}{)}$, $\underset{\sim}{(}\,\alpha \underset{\sim}{\cup} \beta\,\underset{\sim}{)}$, $\underset{\sim}{\neg}\underset{\sim}{(}\,\alpha\,\underset{\sim}{)}$, and $\underset{\sim}{\gamma}\underset{\sim}{(}\,\alpha\,\underset{\sim}{)}$, are γ-expressions over Σ, and $L(\underset{\sim}{(}\,\alpha \underset{\sim}{\cdot} \beta\,\underset{\sim}{)}) = L(\alpha)\cdot L(\beta)$, $L(\underset{\sim}{(}\,\alpha \underset{\sim}{\cup} \beta\,\underset{\sim}{)}) = L(\alpha) \cup L(\beta)$,

   $L(\underset{\sim}{\neg}\underset{\sim}{(}\,\alpha\,\underset{\sim}{)}) = \Sigma^* - L(\alpha)$, and $L(\underset{\sim}{\gamma}\underset{\sim}{(}\,\alpha\,\underset{\sim}{)}) = \gamma(L(\alpha))$,
3) That's all.

Having thus made clear the distinction between a γ-expression α and the set $L(\alpha)$ it defines, we will frequently ignore the distinction when there can be no confusion. Thus we write $\Sigma^* = \sigma \cup \neg(\sigma)$ instead of $\Sigma^* = L(\,\underset{\sim}{(}\,\sigma \underset{\sim}{\cup} \underset{\sim}{\neg}\underset{\sim}{(}\,\sigma\,\underset{\sim}{)}\,\underset{\sim}{)}\,)$. Similarly, for any set of letters $V \subset \Sigma$,

$$V^* = \neg(\Sigma^* \cdot (\Sigma - V)\cdot\Sigma^*)$$

since V^* consists precisely of those words in Σ^* which do not contain a symbol not in V. Thus there is a γ-expression α over Σ such that $L(\alpha) =$

V^*. DeMorgan's law gives us intersection, and then the identities

$$V^n = \Sigma^n \cap V^*, \text{ and}$$
$$\Sigma^n = \gamma(V^n)$$

imply that from a γ-expression of length s for Σ^n we can obtain a γ-expression of length s + c for V^n, and conversely from V^n to Σ^n, for some constant c and all s, $n \in N$. We shall show below that in general s may be much smaller than n.

<u>Definition</u>: <u>Empty</u> $(\Sigma) = \{\alpha \mid \alpha$ is a γ-expression over Σ and $L(\alpha) = \emptyset\}$.

Since the regular (finite automaton recognizable) subsets of Σ^* are closed under $\cdot$, $\cup$, $\neg$, and γ, it follows that <u>Empty</u>(Σ) is recursive and in fact primitive recursive. One simply constructs a finite automaton for $L(\alpha)$ and tests whether the automaton accepts some word; there are well-known procedures to do this. <u>A priori</u> analysis of this procedure however indicates that from <u>deterministic</u> automata for γ-expressions α, β one would obtain a <u>non-deterministic</u> automaton for $\alpha \cdot \beta$ or $\gamma(\alpha)$, and then would have to apply the "subset construction" of Rabin-Scott [13] to obtain an automaton for $\neg(\alpha \cdot \beta)$ or $\neg\gamma(\alpha)$. Since the subset construction can exponentially increase the number of states in the automaton, γ-expressions in which k complementations alternated with γ's and concatenations can lead to an automaton with $t_k(2)$ states. The time and space required by a Turing machine which recognizes <u>Empty</u> (Σ) by the procedure outlined above <u>can</u> be bounded above by

$$t_n(c) = 2^{2^{\cdot^{\cdot^{\cdot^{2^c}}}}} \Big\} n$$

for some constant c and all γ-expressions of length n ≥ 0. It will follow from results below that such absurd inefficiency is inevitable.

Definition: A Turing machine $\mathfrak{M}$ recognizes a set $A \subset \Sigma^*$ if, when started with any word $x \in \Sigma^*$ on its tape, $\mathfrak{M}$ halts in a designated accepting state iff $x \in A$.

Let $f: N \to N$. The space complexity of a set $A \subset \Sigma^*$ is at most f almost everywhere, written

$$\text{SPACE}(A) \le f \text{ (a.e.)}$$

iff there is a Turing machine which recognizes A and which, for all but finitely many $x \in A$, uses at most $f(\ell(x))$ tape squares in its computation on input x. The space complexity of A exceeds f infinitely often written

$$\text{SPACE}(A) > f \text{ (i.o.)}$$

iff it is not true that SPACE (A) ≤ f (a.e.).

We shall use Turing's original one tape, one read-write head model of Turing machine, and define the number of tape squares used during the computation on input x to be the larger of $\ell(x)$ and the number of tape squares visited by the read-write head. Then by convention at least $\max\{\ell(x), 1\}$ tape squares are used in a computation on any input word x.

We briefly review some well-known facts, first established by Stearns, Hartmanis, and Lewis [15], about space-bounded Turing machine computations.

<u>Definition</u>: A function $f\colon N \to N$ is <u>tape constructible</u> iff there is a Turing machine which, started with any input word of length $n \geq 0$, halts having used exactly $f(n)$ tape squares.

<u>Fact 1</u>: $t_0 + 1 = \lambda n[n+1]$ is tape constructible. For any $k > 0$, t_k is tape constructible.

<u>Fact 2</u>: If $f\colon N \to N$ is tape constructible, and SPACE (A) $\leq f$ (a.e.) for some $A \subset \Sigma^*$, then there is a Turing machine which recognizes A which halts on <u>every</u> input $x \in \Sigma^*$ using at most $f(\ell(x))$ tape squares.

Hence, SPACE (A) $\leq f \Leftrightarrow$ SPACE $(\Sigma^*-A) \leq f$.

<u>Fact 3</u>: If $f\colon N \to N$ is tape constructible, then there is an $A \subset \{0,1\}^*$ such that

$$\text{SPACE (A)} \leq f \text{ and}$$

$$\text{SPACE (A)} > g \quad \text{(i.o.)}.$$

for any $g\colon N \to N$ such that

$$\lim_{n \to \infty} \frac{g(n)}{f(n)} = 0.$$

Our proof consists of a sequence of reductions of one decision or recognition problem to another. In contrast to the usual reductions of recursive function theory, our reductions must be computationally efficient. We introduce a particular notion of efficient reduction which is sufficient for our purposes.

<u>Definition</u>: Let Σ_1, Σ_2 be finite sets of symbols, and $A_1 \subset \Sigma_1^*$, $A_2 \subset \Sigma_2^*$. A_1 is efficiently reducible to A_2, written

$$A_1 \;\underline{\text{eff}}\; A_2$$

providing there is a polynomial p and a Turing machine which, started with any word $x \in \Sigma_1^*$ on its tape, eventually halts with a word $y \in \Sigma_2^*$ on its tape such that

1) $x \in A_1 \Leftrightarrow y \in A_2$, and
2) the number of tape squares used in the computation on input x is at most $p(\ell(x))$ (and <u>a fortiori</u> $\ell(y) \le p(\ell(x))$).

We remark that all the reductions which are described below require only a linear polynomial number of tape squares and a polynomial number of steps, but to minimize the demands on the readers intuition (since we never actually give a flow-chart or table of quadruples for the Turing machines we describe) we allow polynomials of any degree. Even so, <u>eff</u> is much more restricted than is necessary to prove Theorem 2.

<u>Fact 4</u>. <u>eff</u> is a transitive relation on sets of words.

<u>Fact 5</u>. If A_1 <u>eff</u> A_2 and SPACE $(A_2) \le f$ (a.e.), then there is a polynomial p such that

$$\text{SPACE } (A_1) \le \lambda n[\max\{f(m) \mid m \le p(n)\} + p(n)] \text{ (a.e.)}$$

<u>Fact 6</u>. If A_1 eff A_2 and SPACE $(A_1) > t_{k+1}$ (i.o.), then SPACE $(A_2) > t_k$ (i.o.).

Proof. Immediate from Fact 5 and that observation that for any polynomial p, $t_k(p(n)) + p(n) \le t_{k+1}(n)$ for all sufficiently large n.

The proof of Theorem 2 can now be summarized.

<u>Proof of Theorem 2</u>: We will establish below that

<u>Empty</u> ({0,1}) <u>eff</u> <u>WSIS</u>

<u>Empty</u> (Σ) <u>eff</u> <u>Empty</u> ({0,1}) for any finite Σ,

and finally that for any k and for any set $A \subset \{0,1\}^*$ such that SPACE (A) $\le t_k$ (a.e.) there is a finite Σ such that

A eff Empty (Σ)

From fact 4, we have A eff WSIS for any A and k such that SPACE (A) $\le t_k$ (a.e.).

Then from facts 1, 3 and 6 we conclude that SPACE (WSIS) $> t_{k-1}$(i.o.) for any k. Q.E.D.

It remains only to establish the required reductions.

Lemma 1: Empty($\{0,1\}$) eff WSIS

Proof: For any γ-expression α over $\{0,1\}$ we shall show how to construct a formula $F_\alpha \in L_{SIS}$ with two free integer variables and one free set variable. For any set $M \subset N$, let $C_M: N \to \{0,1\}$ be the characteristic function of M, that is, $C_M(n) = 1 \Leftrightarrow n \in M$. The formula F_α will be constructed so that for n, m $\in$ N, $M \subset N$, M finite:

$$F_\alpha(n,m,M) \text{ is true} \Leftrightarrow [[n < m \text{ and } C_M(n)\cdot C_M(n+1) \dots C_M(m-1) \in L(\alpha)] \text{ or } [n = m \text{ and } \lambda \in L(\alpha)]].$$

F_α is constructed by induction on the definition of γ-expressions. If α is 0 or 1, then

$F_0(x,y,X)$ is $[y = x+1$ and $\neg (x \in X)]$,

$F_1(x,y,X)$ is $[y = x+1$ and $x \in X]$.

If α is $(\beta \cdot \delta)$, then

$F_\alpha(x,y,X)$ is $(\exists z)[x \le z$ and $z \le y$ and $F_\beta(x,z,X)$ and $F_\delta(z,y,X)]$.

If α is $\underset{\sim}{\gamma}\,\underset{\sim}{(}\,\beta\,\underset{\sim}{)}$, then

$F_\alpha(x,y,X)$ is $(\exists X_0)[F_\beta(x,y,X_0)]$.

If α is $\underset{\sim}{(}\,\beta \underset{\sim}{\cup} \delta\,\underset{\sim}{)}$ or $\underset{\sim}{\neg}\,\underset{\sim}{(}\,\beta\,\underset{\sim}{)}$, then F_α is $[F_\beta$ <u>or</u> $F_\delta]$ or $[x \leq y$ <u>and</u> $\neg F_\beta(x,y,X)]$, respectively.

It is clear that there is a Turing machine which, given an input $\alpha \in \{0, 1, \underset{\sim}{(}, \underset{\sim}{)}, \underset{\sim}{\cup}, \underset{\sim}{\gamma}, \underset{\sim}{\cdot}, \underset{\sim}{\neg}\}^*$, can test whether α is a well-formed γ-expression and, if so, print out the sentence

$$\neg(\exists x)(\exists y)(\exists X)[F_\alpha(x,y,X)],$$

never using more space than some fixed polynomial in $\ell(\alpha)$. (If α is not well-formed, the machine prints out $(\exists x)[x = x+1]$.) Hence, <u>Empty</u>$(\{0,1\})$ <u>eff</u> <u>WSIS</u>. Q.E.D.

It will be convenient to work with larger symbol sets than $\{0,1\}$, but a trivial coding will demonstrate that this involves no loss of generality.

Let Σ be any finite set of symbols with $|\Sigma| \geq 2$. Say $0, 1 \in \Sigma$, $0 \neq 1$. Then for any $n \geq 1$, there is a γ-expression over Σ for $(\Sigma^n)^*$. To see this, consider a word in Σ^* <u>not</u> in $(0^{n-1}\ 1)^*$. Such a word either fails to begin with $0^{n-1}\ 1$, fails to end with 1, or contains a subword in $0\ \Sigma^{n-1}(\Sigma\text{-}0)$ or $1\ \Sigma^{n-1}(\Sigma\text{-}1)$. Hence

$$\lambda \cup \neg((0^{n-1}1)^*) = \neg(0^{n-1}1\ \Sigma^*) \cup \neg(\Sigma^*\ 1) \cup (\Sigma^*\ 0\ \Sigma^{n-1}(\Sigma\text{-}0)\ \Sigma^*)$$
$$\cup(\Sigma^*\ 1\ \Sigma^{n-1}(\Sigma\text{-}1)\ \Sigma^*),$$

and so, noting that $\lambda = \neg(\Sigma\cdot\Sigma^*)$, we have

$$(0^{n-1}\ 1)^* = \neg(\lambda \cup \neg((0^{n-1}\ 1)^*)) \cup \lambda,$$

and

$$(\Sigma^n)^* = \gamma((0^{n-1}\ 1)^*).$$

Now given any finite set Σ_1 choose n sufficiently large that $|\Sigma^n| \geq |\Sigma_1|$ and let h: $\Sigma_1 \to \Sigma^n$ be any one-one function. Extend h to a one-one map from $P(\Sigma_1^*)$ into $P((\Sigma^n)^*)$ by the obvious rules $h(\lambda) = \lambda$, $h(x\sigma_1) = h(x) \cdot h(\sigma_1)$ for $x \in \Sigma_1^*$, $\sigma_1 \in \Sigma_1$, and $h(A) = \bigcup_{x \in A} \{h(x)\}$ for $A \subset \Sigma_1^*$. There is then a γ-expression over Σ for $h(\Sigma_1^*)$, because a word fails to be in $h(\Sigma_1^*)$ either because its length is not a multiple of n, or else because it contains a subword of length n not in $h(\Sigma_1)$ which begins at a position congruent to one modulo n:

$$\Sigma^* - h(\Sigma_1^*) = \neg((\Sigma^n)^*) \cup (\Sigma^n)^* \cdot (\Sigma^n - h(\Sigma_1)) \cdot (\Sigma^n)^*.$$

Lemma 2: (Coding) Let Σ_1, Σ be finite sets of symbols with $|\Sigma| \geq 2$. Let h: $P(\Sigma_1^*) \to P((\Sigma^n)^*)$ be the extension of a one-one function from Σ_1 to Σ^n for some $n \geq 1$. There is a Turing machine which, started with a γ-expression α over Σ_1, halts with a γ-expression β over Σ on its tape such that

$$h(L(\alpha)) = L(\beta).$$

Moreover the space used during the computation with input α is bounded by a polynomial in $\ell(\alpha)$.

Proof. The transformation of α to β operates by applying the following rules recursively.

If $\alpha \in \Sigma_1$, β is set equal to an expression for $h(L(\alpha))$.

If α is $\underset{\sim}{(} \alpha_1 \underset{\sim}{\cdot} \alpha_2 \underset{\sim}{)}$ or $\underset{\sim}{(} \alpha_1 \underset{\sim}{\cup} \alpha_2 \underset{\sim}{)}$, then β is $\underset{\sim}{(} \beta_1 \underset{\sim}{\cdot} \beta_2 \underset{\sim}{)}$ or $\underset{\sim}{(} \beta_1 \underset{\sim}{\cup} \beta_2 \underset{\sim}{)}$, respectively, where β_1, β_2 are the transforms of α_1, α_2.

If α is $\underset{\sim}{\gamma} \underset{\sim}{(} \alpha_1 \underset{\sim}{)}$, then β is

$$\underset{\sim}{\neg}\underset{\sim}{(}\underset{\sim}{(}\underset{\sim}{\neg}\underset{\sim}{(}\underset{\sim}{\gamma}\underset{\sim}{(}\beta_1\underset{\sim}{)}\underset{\sim}{)}\underset{\sim}{\cup}\beta_{\Sigma_1}\underset{\sim}{)}\underset{\sim}{)}$$

where β_1 is the transform of α_1 and β_{Σ_1} is a γ-expression over Σ for $\Sigma^*-h(\Sigma_1^*)$. (Note that $h(\gamma_{\Sigma_1}(A)) = \gamma_{\Sigma}(h(A)) \cap h(\Sigma_1^*)$ for $A \subset \Sigma_1^*$, which justifies this rule.)

Finally, if α is $\underset{\sim}{\neg}\underset{\sim}{(}\alpha_1\underset{\sim}{)}$, then β is $\underset{\sim}{\neg}\underset{\sim}{(}\underset{\sim}{(}\beta_1 \underset{\sim}{\cup} \beta_{\Sigma_1}\underset{\sim}{)}\underset{\sim}{)}$, since

$h(\Sigma_1^*-A) = h(\Sigma_1^*) - h(A) = \Sigma^* - (h(A) \cup (\Sigma^*-h(\Sigma_1^*)))$ for $A \subset \Sigma_1^*$.

It is clear that a Turing machine can carry out this recursive transformation within the required space bound. Q.E.D.

<u>Corollary</u>: <u>Empty</u> (Σ) <u>eff</u> <u>Empty</u> $\{0,1\}$ for any finite Σ.

<u>Proof</u>: Code Σ into $\{0,1\}$ via h as in Lemma 2. Then $\alpha \in$ <u>Empty</u> $(\Sigma) \Leftrightarrow$ $L(\alpha) = \emptyset \Leftrightarrow h(L(\alpha)) = \emptyset \Leftrightarrow L(\beta) = \emptyset \Leftrightarrow \beta \in$ <u>Empty</u> $\{0,1\}$. Q.E.D.

We now show how, given a γ-expression for Σ^n, one can construct a γ-expression of about the same size describing any desired computation of a Turing machine, providing the states and symbols of the Turing machine can be represented in Σ and the computation only requires n tape squares. This construction will be applied recursively to obtain γ-expressions of size n for $\Sigma^{t_k(n)}$, and will then finally be used to conclude that A <u>eff</u> <u>Empty</u> (Σ) for any $A \subset \{0,1\}^*$ such that SPACE $(A) \leq t_k$ (a.e.).

Definition: Let $\mathfrak{M}$ be any Turing machine with tape symbols T and states S. Assume $b \in T$ where "b" designates a blank tape square. An instantaneous description (i.d.) of $\mathfrak{M}$ is a word in $(T \cup (S \times T))^{\dagger}$ which contains exactly one symbol in $S \times T$. Given any i.d. $x = y\cdot(s,t)\cdot z$ for $y, z \in T^*$, $s \in S$, $t \in T$, the next i.d., $\text{Next}_{\mathfrak{M}}(x)$ is defined as follows: if when $\mathfrak{M}$ is in state s with its read-write head scanning symbol t, $\mathfrak{M}$ enters state s' and writes symbol $t' \in T$, then $\underline{\text{Next}_{\mathfrak{M}}}(x)$ is

$y\cdot(s', t')\cdot z$	if $\mathfrak{M}$ does not shift its head,
$y\cdot t'(s', u)\cdot w$	if $\mathfrak{M}$ shifts its head right and $z = uw$ for $u \in T$, $w \in T^*$,
$w\cdot(s', u)\cdot t'\cdot z$	if $\mathfrak{M}$ shifts its head left and $y = wu$ for $u \in T$, $w \in T^*$.

$\text{Next}_{\mathfrak{M}}(x)$ is undefined if (s,t) is a halting condition, or if (s,t) is the rightmost (leftmost) symbol of x and $\mathfrak{M}$ shifts right (left). Let $\underline{\text{Next}_{\mathfrak{M}}}(x,0) = x$ if x is an i.d., undefined otherwise; $\underline{\text{Next}_{\mathfrak{M}}}(x, n+1) = \text{Next}_{\mathfrak{M}}(\text{Next}_{\mathfrak{M}}(x,n))$.

Finally, let # be a symbol not in $T \cup (S \times T)$. The computation $\text{Comp}(\mathfrak{M},x)$ of $\mathfrak{M}$ from x is singleton set consisting of the following word in $(\{\#\} \cup T \cup (S \times T))^*$:

$$\text{Comp}(\mathfrak{M},x) = \{\#\cdot\text{Next}_{\mathfrak{M}}(x,0)\cdot\#\cdot\text{Next}_{\mathfrak{M}}(x,1)\cdot\#\cdot\cdots\cdot\#\cdot\text{Next}_{\mathfrak{M}}(x,n)\cdot\#\}$$

---

† $S \times T = \{(s,t) \mid s \in s \text{ and } t \in T\}$. We assume $T \cap (S \times T) = \emptyset$

where n is the least integer such that (q_a,t) occurs in $\mathrm{Next}_{\mathfrak{M}}(x,n)$ for some $t \in T$ and designated halting state q_a. $\mathrm{Comp}(\mathfrak{M},x) = \phi$ if there is no such n.

Remark: Note that our definition of computation differs from the one commonly in the literature. The computation $\mathrm{Comp}(\mathfrak{M},x)$ is defined for i.d.'s x, not input words x. Moreover, all i.d.'s in $\mathrm{Comp}(\mathfrak{M},x)$ have exactly the same length. A key property of $\mathrm{Comp}(\mathfrak{M},x)$ is given next.

Fact 7: Given $\mathfrak{M}$ as in the preceding definition, let $\Sigma = \{\#\} \cup T \cup (S \times T)$. Then for any i.d. $y \in \Sigma^*$, the $n\text{-}1^{st}$, n^{th} and $n{+}1^{st}$ symbols of y uniquely determine the nth symbol of $\mathrm{Next}_{\mathfrak{M}}(y)$ for $1 < n < \ell(y)$ providing $\mathrm{Next}_{\mathfrak{M}}(y)$ is defined.

Hence, there is a partial function $f_{\mathfrak{M}}: \Sigma^3 \to \Sigma$ such that if $\sigma_1,\sigma_2,\sigma_3$ are the $n\text{-}1^{st}$, n^{th}, $n{+}1^{st}$ symbols of $\mathrm{Comp}(\mathfrak{M},x)$, then $f_{\mathfrak{M}}(\sigma_1,\sigma_2,\sigma_3)$ is the $n+\ell(x)+1^{st}$ symbol of $\mathrm{Comp}(\mathfrak{M},x)$ for $1 < n < \ell(\mathrm{Comp}(\mathfrak{M},x))-\ell(x)$ and any i.d. x such that $\mathrm{Comp}(\mathfrak{M},x) \neq \phi$. Also, $f_{\mathfrak{M}}(\sigma_1,\sigma_2,\sigma_3) = \phi$, if $\sigma_2 \in (S \times T)$ and σ_2 is a halting condition of $\mathfrak{M}$.

Lemma 3: (Simulation) Let $\mathfrak{M}$ be a Turing machine with states S, symbols T, and designated halting state $q_a \in S$. Let $\Sigma = \{\#\} \cup T \cup (S \times T)$. There is a Turing machine $\mathcal{F}(\mathfrak{M})$ which, started with any word $y \cdot \# \cdot \alpha$ on its tape where y is an i.d. of $\mathfrak{M}$ and α is a γ-expression over Σ such that $L(\alpha) = \Sigma^n$ for some $n > 0$, halts with a γ-expression β over Σ such that

$$L(\beta) = \mathrm{Comp}(\mathfrak{M}, b^n \cdot y \cdot b^n).$$

Moreover, there is a polynomial p such that $\mathcal{F}(\mathfrak{M})$ never uses more than $p(\ell(y\cdot\#\cdot\alpha))$ tape squares in its computation.

Proof: We shall describe how to construct the γ-expression β for $\mathrm{Comp}(\mathfrak{M}, b^n y b^n)$ from $y\cdot\#\cdot\alpha$ where $L(\alpha) = \Sigma^n$. We begin by noting that the words in Σ^* not equal to $\mathrm{Comp}(\mathfrak{M}, b^n y b^n)$, i.e., $\neg(\mathrm{Comp}(\mathfrak{M}, b^n y b^n))$, can be characterized as follows:

1) words that do not begin with $\# b^n y b^n \#$, or
2) words that do not contain q_a, or
3) words that do not end with #, or
4) words that violate the functional condition determined by $f_{\mathfrak{M}}$ in Fact 7.

These four sets of words can also be described by the formulas

1') $\neg(\#\cdot(L(\alpha) \cap b^*)\cdot y\cdot(L(\alpha) \cap b^*)\cdot\#\cdot\Sigma^*)$,

2') $\neg(\Sigma^*\cdot(\{q_a\} \times T)\cdot\Sigma^*)$,

3') $\neg(\Sigma^*\cdot\#)$,

4') $\bigcup_{\sigma_1,\sigma_2,\sigma_3 \in \Sigma} [\Sigma^*\cdot\sigma_1\sigma_2\sigma_3\cdot L(\alpha)\cdot\Sigma^{\ell(y)-1}\cdot L(\alpha)\cdot(\Sigma - f_{\mathfrak{M}}(\sigma_1,\sigma_2,\sigma_3))\cdot\Sigma^*]$

But it is easy to see how to construct γ-expressions directly from (1')-(4'), and therefore β is simply the complement of the union of these four expressions. Note that $\ell(\beta) \le c\cdot\ell(y\#\alpha)$ for some constant c which depends only on $\mathfrak{M}$, and not on y or α. Moreover a Turing machine $\mathcal{F}(\mathfrak{M})$ which constructs β from $y\#\alpha$ need never use more tape squares than $\ell(\beta)$, and so certainly runs within a polynomial space bound. Q.E.D.

Definition: A Σ-t_k-TM is a Turing machine such that for some polynomial p, some function $f_k \geq t_k$, and all $n > 0$, when the Turing machine is started with 0^n on its tape, it halts with a word α on its tape such that

1) α is a γ-expression over Σ and $L(\alpha) = \Sigma^{f_k(n)}$,

2) the number of tape squares used in the computation is at most p(n).

Lemma 4: If there is a Σ'-t_k-TM for any finite Σ', then there is a Σ-t_k-TM for any Σ such that $|\Sigma| \geq 2$.

Proof: Code Σ' into Σ as in Lemma 2. Details are left to the reader.

Q.E.D.

Lemma 5: For any $k \geq 0$ and any Σ with $|\Sigma| \geq 2$, there is a Σ-t_k-TM.

Proof: A Σ-t_0-TM simply prints an expression for $\gamma(\sigma^n)$ from input 0^n, where $\sigma \in \Sigma$ is any symbol. Proceeding by induction, assume there is a Σ-t_k-TM. Let $\mathfrak{M}_k$ be a Turing machine which, started with 0^n on its tape for any $n > 0$, lays out $t_k(n)$ tape squares on its tape and then uses these tape squares to cycle through some number $f_{k+1}(n) \geq 2^{t_k(n)} = t_{k+1}(n)$ steps before finally halting. Since t_k is tape-constructible, it is easy to obtain $\mathfrak{M}_k$ as described. Choose Σ as in the simulation lemma applied to $\mathfrak{M}_k$.

The Σ-t_{k+1}-TM operates as follows: Given 0^n, use the Σ-t_k-TM to obtain α such that $L(\alpha) = \Sigma^{f_k(n)}$. Apply $\mathcal{F}(\mathfrak{M}_k)$ of the simulation lemma to $(q_0,0)0^{n-1}\cdot\#\cdot\alpha$ where q_0 is the start state of $\mathfrak{M}_k$. This yields a

γ-expression β such that $L(\beta) = \text{Comp}(\mathfrak{M}_k, x)$ where $x = b^{f_k(n)} \cdot (q_0, 0)0^{n-1} \cdot b^{f_k(n)}$.

But $\text{Comp}(\mathfrak{M}_k, x)$ is defined since $\mathfrak{M}_k$ halts on input 0^n within $t_k(n) \leq f_k(n)$ tape squares. Moreover, $\ell(\text{Comp}(\mathfrak{M}_k, x)) \geq t_{k+1}(n)$ since $\mathfrak{M}_k$ runs for at least $t_{k+1}(n)$ steps. Hence, the output of the Σ-t_{k+1}-TM is simply $\gamma(\beta)$.

Since by hypothesis α is obtainable in space $p_1(n)$ for some polynomial p_1, and similarly β is obtainable in space $p_2(n+1 + p_1(n))$ for some polynomial p_2, the entire process requires only polynomial space. Q.E.D.

<u>Lemma 6</u>: For any set $A \subset \{0,1\}^*$, if $\text{Comp}(A) \leq t_k$ (a.e.) for some $k \geq 0$, then there is a finite Σ such that A <u>eff</u> <u>Empty</u>(Σ).

<u>Proof</u>: Let $\mathfrak{M}$ be a Turing machine which recognizes $\{0,1\}^* - A$ and for every $x \in \{0,1\}^*$, $\mathfrak{M}$ halts using at most $t_k(\ell(x))$ tape squares. By Fact 2, there is such an $\mathfrak{M}$.

Choose Σ as in the simulation lemma applied to $\mathfrak{M}$.

The Turing machine which efficiently reduces A to <u>Empty</u> (Σ) operates as follows: given $x \in \{0,1\}^*$, use a Σ-t_k-TM to obtain a γ-expression α such that $L(\alpha) = \Sigma^{f_k(n)}$ for $n = \ell(x)$. Apply $\mathcal{F}(\mathfrak{M})$ of the simulation lemma to $(q_0, u) \cdot w \cdot \# \cdot \alpha$ where q_0 is the start state of $\mathfrak{M}$, and $x = uw$ for $u \in \{0,1\}$, $w \in \{0,1\}^*$. (We ignore the case $x = \lambda$.) This yields a γ-expression β which we claim is the desired output.

Since $\mathfrak{M}$ requires space at most $t_k(n)$, we conclude that $\text{Comp}(\mathfrak{M}, y)$ where $y = b^{f_k(n)} \cdot (q_0, u) \cdot w \cdot b^{f_k(n)}$ is nonempty iff x is accepted by $\mathfrak{M}$. Hence $x \in A \Leftrightarrow x$ is not accepted by $\mathfrak{M} \Leftrightarrow \text{Comp}(\mathfrak{M}, y) = \emptyset \Leftrightarrow L(\beta) = \emptyset \Leftrightarrow \beta \in$ <u>Empty</u>(Σ). This verifies our claim that β is a correct output.

As in the preceding lemma, the Turing machine transforming x to β requires space at most a polynomial in $\ell(x)$. Q.E.D.

This completes the lemmas required for Theorem 2.

It is not hard to extend this argument to obtain Theorem 1. We use a stronger form of Fact 3 due to Blum [1] to obtain from the proof of Theorem 2 more information about the frequency of the (i.o.) condition in the statement that Comp(<u>WSIS</u>) > t_k (i.o.).

<u>Theorem 3</u>: The following decidable full and weak second order theories are not elementary-recursive: two successors, countable linear order, countable well-order, unary function with countable domain, unit interval under ≤. Also, first order theory of two successors with length and prefix predicates, and the first order theory of <N,+,P>, where P(x,y) ≡ [x is a power of two and x divides y], are decidable but not elementary.[†]

These results follow by reasonably straightforward efficient reductions of <u>WSIS</u> to each of these theories.

γ-expressions are themselves of interest as a decidable but non-elementary word problem.

<u>Corollary:</u> <u>Empty</u>({0,1}) is not elementary-recursive.

<u>Further</u> <u>remarks</u>:

(1). The results and methods described here were developed in May, 1972. [9] This paper is a revised version of a preliminary report with the same title written at that time. Since then, in collaboration with

---

† The decidability of these theories is shown in [6,12].

M.J. Fischer, M.O. Rabin, and L. Stockmeyer, J. Ferrante and C. Rackoff, close upper and lower bounds on space or time have been obtained for most of the classical decidable theories in logic as well as for various notations related to γ-expressions.

Some of the more interesting results to appear in forthcoming papers are

(i) (Meyer) The satisfiability problem for sentences in the first order theory of linear order is not elementary; in fact space $t_{\epsilon\cdot n}(n)$ is required for some $\epsilon > 0$. <u>WSIS</u> also requires this much space. An upper bound $t_{c\cdot n}(n)$ follows from Rabin's proof that S 2 S is decidable [12].†

(ii) (Stockmeyer) The emptiness problem for expressions involving only the operation of $\cup$, $\cdot$, $\neg$ is not elementary, that is, the γ-operation is unnecessary. The simulation lemma and its proof become considerably more subtle.

(iii) (Fischer-Rabin) Any decision procedure for the first-order theory of $\langle N,+\rangle$, that is, Presburger's arithmetic, requires $t_2(\epsilon\cdot n)$ steps even on nondeterministic Turing machines. Ferrante and Rackoff[7], following Cooper[5] and Oppen[11], have established an upper bound of space $t_2(\epsilon\cdot n)$.

(iv) (Fischer-Rabin) Any decision procedure for the first order theory of $\langle N-\{0\},\cdot\rangle$ requires time $t_3(\epsilon\cdot n)$ even on nondeterministic Turing machines. Rackoff has shown that space $t_3(c\cdot n)$ is sufficient.

---

† In [12], Rabin inaccurately claims his decision procedure is elementary. In a personal communication, he has informed me that he was aware that his procedure required space $t_{c\cdot n}(n)$, but that he misunderstood the definition of elementary.

(v) (Fischer) Let $\mathcal{S}$ be any class of structures with a binary associative operator * and the property that for arbitrarily large n there exists $s \in S \in \mathcal{S}$ such that

$$s^n \neq s^m \text{ for } 1 \leq m < n,$$

where $s^m = \underbrace{s * s * \cdots * s}_{m}$. Then any decision procedure for satisfiability over $\mathcal{S}$ of sentences in the first order language of * requires $t_1(\epsilon \cdot n)$ steps. This general result applies to nearly all the familiar decidable theories in logic, except for the propositional calculus and pure equality.

(vi) (Meyer) The decision problem for satisfiability of sentences in monadic predicate calculus with only seven (approximately) quantifiers requires time $t_1(\epsilon \cdot n)$ even on nondeterministic Turing machines; time $t_1(c \cdot n)$ is achievable on nondeterministic Turing machines.

(vii) (Fischer-Meyer) The decision problem for satisfiability of sentences in the first order language of a single monadic function is not elementary.

(2) Abstract complexity theory has been open to the criticism of being unable to exhibit "natural" decision problems in which phenomena such as speed-up appeared. Applying Blum's results [2] on effective speed-up to our simulation of Turing machines via <u>WSIS</u>, we can show that given <u>any</u> decision procedure for <u>WSIS</u>, one can <u>effectively</u> construct a new decision procedure for <u>WSIS</u> which is much faster (faster by t_k for any k) than the given procedure on at least one

sentence of length n for all sufficiently large integers n. Similar results apply to the other decision procedures mentioned above.

(3) The relation <u>eff</u> can be characterized in a manner similar to the definition of the elementary functions or the primitive recursive functions. $\varepsilon^{2.5}$, so called because it lies properly between the Grzegorczyk classes ε^2 and ε^3, is defined inductively as follows:

1. $x \dot{-} y$, $x+y$, $x \cdot y$, $x^{\lfloor \log_2 y \rfloor} \in \varepsilon^{2.5}$,
2. $\varepsilon^{2.5}$ is closed under explicit transformation (substituting constants and renaming or identifying variables),
3. $\varepsilon^{2.5}$ is closed under composition of functions, and
4. $\varepsilon^{2.5}$ is closed under limited recursion, limited sum and limited minimization.†
5. That's all.

If we identify words in Σ^* with the integers they represent in $|\Sigma|$-adic notation, and for any set $A \subset \Sigma^*$ let $C_A: N \to \{0,1\}$ be the characteristic function of the set of integers identified with A, then B <u>eff</u> A if and only if $C_A(x) = C_B(f(x))$ for some $f \in \varepsilon^{2.5}$ and all $x \in N$.

Essentially $\varepsilon^{2.5}$ provides a high-level programming language in which one can formally express the procedures we informally claimed could be carried out by polynomial space-bounded Turing machines. In this manner our proof could be presented in a completely formal fashion without appeal to intuition about the space requirements of computations. We prefer the latter approach.

---

†See Grzegorczyk's paper for definitions. [8]. Closure under limited recursion actually implies closure under limited sum and limited minimization.

Acknowledgments: Larry Stockmeyer's proof, that any problem decidable in nondeterministic polynomial time is deterministic polynomial time reducible to the regular expressions not equal to Σ^*, provided the key idea of the simulation lemma. Jeanne Ferrante and Charles Rackoff worked out the efficient reductions of WSIS mentioned in Theorem 3. Patrick Fischer correctly suggested that the use of * in my original proof was inessential.

My colleague Michaël J. Fischer's suggestions and attention were extremely helpful, as they invariably have been in the past.

October, 1973
Cambridge, Mass.

REFERENCES

1. Blum, M. A machine-independent theory of the complexity of recursive functions, Jour. Assoc. Comp. Mach., 14, 2 (April, 1967), 322-336.

2. Blum, M. On effective procedures for speeding up algorithms, Jour. Assoc. Comp. Mach., 18, 2 (April, 1971), 290-305.

3. Büchi, J.R. and C.C. Elgot, Decision problems of weak second order arithmetics and finite automata, Part I, (abstract), AMS Notices, 5 (1959), 834.

4. Büchi, J.R. Weak second order arithmetic and finite automata, Zeit. f. Math. Log. and Grund. der Math., 6 (1960), 66-92.

5. Cooper, D.C. Theorem-proving in arithmetic without multiplication, Computer and Logic Group Memo. No. 16, U.C. of Swansea, April, 1972, to appear in Machine Intelligence 7.

6. Elgot, C.C. and M.O. Rabin, Decidability and undecidability of extensions of second (first) order theory of (generalized) successor, Jour. Symb. Logic, 31, 2 (June, 1966), 169-181.

7. Ferrante, J. and C. Rackoff, A decision procedure for the first order theory of real addition with order, Project MAC Tech. Memo 33, Mass. Inst. of Technology (May, 1973), 16pp., to appear SIAM Jour. Comp.

8. Grzegorczyk, A. Some classes of recursive functions, Rozprawy Matematyczne, 4 (1953), Warsaw, 1-45.

9. Meyer, A.R. Weak SIS cannot be decided (abstract 72T-E67), AMS Notices, 19, 5 (August, 1972), p. A-598.

10. Meyer, A.R. and L.J. Stockmeyer, The equivalence problem for regular expressions with squaring requires exponential space, 13th Switching and Automata Theory Symp. (Oct. 1972), IEEE, 125-129.

11. Oppen, D.C. Elementary bounds for Presburger arithmetic, 5th ACM Symp. Theory of Computing (April, 1973), 34-37.

12. Rabin, M.O. Decidability of second-order theories and automata on infinite trees, Trans. AMS, 141 (July, 1969), 1-35.

13. Rabin, M.O. and D. Scott, Finite automata and their decision problems, IBM Jour. Research and Development, 3 (1959), 115-125.

14. Ritchie, R.W. Classes of predictably computable functions, Trans. AMS, 106 (1963), 139-173.

15. Stearns, R.E., J. Hartmanis, and P.M. Lewis, III, Hierarchies of memory-limited computations, 6th Switching Theory and Logical Design Symp. (1965), IEEE, 179-190.

16. Stockmeyer, L.J. and A.R. Meyer, Word problems requiring exponential time, 5th ACM Symp. Theory of Computing (April, 1973), 1-9.

THE VARIABLE[1]

W.V. Quine

The variable *quâ* variable, the variable *an und für sich* and *par excellence*, is the bindable objectual variable. It is the essence of ontological idiom, the essence of the referential idiom. But it takes some distilling, for it has strong affinities with quite a variety of closely associated notions and devices.

It used to be necessary to warn against the notion of variable numbers, variable quantities, variable objects, and to explain that the variable is purely a notation, admitting only fixed numbers or other fixed objects as its values. This dissociation now seems to be generally understood, so I turn to others.

There is the schematic letter. As recently as 1945, and in as sophisticated a medium as the Journal of Symbolic Logic, I still felt I had to devote three pages to explaining the status of the schematic sentence letters and predicate letters as used in the logic of truth functions and quantification. These letters refer to no objects as values. The sentence letters refer neither to propositions nor to truth values as their values, much less to sentences, and the predicate letters refer neither to properties nor to classes as values, much less to predicates. They are not bindable, they are not objectual, and they do not occur in sentences. They occur in schemata. I devoted three pages to the matter, not with any sense of creativity, but still with a lively sense of warding off basic misunderstandings on the part of most readers. Just five years earlier, in *Mathematical Logic*, I had even thought it unwise to court such misunderstandings by using schematic letters at all. In that book I presented the logic of truth functions and quantification wholly in a metalogical notation of Greek letters and corners. The schematic sentence letters and predicate letters had already been in general use, but the trouble was that their schematic status was seldom clearly appreciated. Now and again they would even get quantified.

A further device that we must take pains to dissociate from the objectual variable is the bindable substitutional variable. The

[1] This work was supported in part by grant GS-2615 of the National Science Foundation.

schematic letter itself of course is already purely substitutional: it does not refer to any objects as values, but merely admits appropriate expressions as substitutes. But besides schematic letters, which are not bindable, there is the use by Ruth Marcus and others of substitutional variables that can be bound by quantifiers and embedded in sentences. These quantifications cannot be read in the objectual way, as meaning that every or some object _x_ of the appropriate category is thus and so. But still they can be explained by clear truth conditions. The universal quantification is true if the sentence under the quantifier comes out true under all substitutions in the appropriate grammatical category; and correspondingly for the existential quantification.

Our quantification over individuals is seen most naturally as objectual. To take it as substitutional would require assuming in our langauge a name, or some uniquely designating expression, for every individual, every creature or particle, however obscure and remote; and this would be artificial at best. When we come to quantify over classes, the substitutional version does look tempting; however, a problem arises.[2] Consider the law:

$$(\underline{W})(\underline{W} \text{ has members } .\supset (\exists \underline{Z})(\underline{Z} \text{ has a member of } \underline{W} \text{ as sole member})).$$

This is gospel for classical set theory. It may be called the law of unit subclasses. But what does it come to when '_W_' and '_Z_' are substitutional? It then requires, of every closed class term that we can write in our language, that if its membership condition happens to be true of any individuals, however obscure and remote, we must be able to write another condition that singles out one of those individuals uniquely. This is as unwelcome as assuming outright in our language a uniquely designating expression for every individual; and if that were welcome, our individual variables could go substitutional too.

Parsons has proposed a modified truth condition for substitutional quantification that averts this effect. For him an existential quantification still counts as true as long as it has a substitution instance that contains free objectual variables and comes out true for some values of them. By this standard the law of unit subclasses is true.

I said that the bindable objectual variable has close associations

---

[2] I am indebted to Oswaldo Chateaubriand and Gilbert Harman for starting me on the following line of thought.

with quite a variety of notions and devices. I have now dissociated it from variable numbers and other variable objects, there being for these no place in the universe. I have dissociated it also from the schematic letter, fond though I am of both the variable and the schematic letter. And I have dissociated it now from the bindable substitutional variable, though recognizing this variable as clear and legitimate as far as it goes.

I have thus been dissociating the bindable objectual variable from associates with which it was apt to be confused. I want next to dissociate it even from quantification, so as to reveal its nature uncolored by those two familiar contexts.

We tend to think of bound variables primarily in quantification. This is because we know how to paraphrase other uses of bound variables into the quantificational use, and because, moreover, there are algorithmic benefits to be gained by doing so. The 'x' of the description '$(\iota x)Fx$' goes over into an 'x' of quantification under Russell's contextual definition of description. The 'x' of the class abstract '$\{x: Fx\}$' goes over into an 'x' of quantification when '$\{x: Fx\}$' is defined as a description '$(\iota y)(x)(x \in y \mathbin{.{\equiv}} Fx)$'. The '$x$' of the functional abstract '$\lambda_x fx$' fares similarly. So does the bound variable of the differential and integral calculus, for it is ultimately a variable of functional abstraction.

Reductions can be made in alternative directions too. Functional abstraction or class abstraction can be taken as basic, and quantification can be defined in terms of either of them. Church took the one line in his lambda calculus, and I the other in my logic based on inclusion and abstraction. Even Peano had a full-blown class abstraction on which he based his existential quantification, though his universal quantification took another line. However, reduction to quantification is generally to be preferred to these alternative reductions, because quantification is wanted not only in set theory but also in elementary theories where there is no call for classes or functions; and moreover the logic of quantification, unlike set theory, is complete, compact, and convenient.

Such, then, is a *standard theory*, in Tarski's phrase; it is simply quantification logic, or the predicate calculus, with one or another fixed lexicon of predicates appropriate to one or another particular subject matter. The ontology of a theory, thus standardized, is the range of values of the variables of quantification; for the variables *are* the variables of quantification. And of course it is clear from the readings of objectual quantification in ordinary language that the ontology consists of those values; for the quantifiers mean 'everything is such that' and 'something is such that'.

Very well, then; where does the dissociation come in? The point I want to make is that the quantitative force of the quantifier, the 'all' and 'some', is irrelevant to the distinctive work of the bound variable and irrelevant to its referential function. The quantitative component needs no variable; it is fully present in the traditional categoricals 'All men are mortal', 'Some Greeks are wise'. Conversely the bound variable is fully taken up with its distinctive work when used in description, class abstraction, functional abstraction, integration, no less than when used in quantification; and yet these other uses connote nothing of 'all' and 'some', except of course as we impose reductive definitions in the direction of quantification.

Minutes ago I extolled that direction of reduction, as in a sense more basic than reduction to class abstraction or functional abstraction. There is, however, a use of the bound variable that is more basic still than its use in quantification. It carries no connotation of 'all' or 'some' or class or function, but shows rather the distinctive work of the bound variable without admixture. This basic and neglected idiom is the relative clause, mathematically regimented as the 'such that' idiom: 'x such that Fx'. It is not a singular term, neither a singular description nor a class abstract; it is a general term, a predicate. It has its use where we have a complex sentence that mentions some object a perhaps midway, perhaps repeatedly, and we want to segregate a complex adjective or common noun that may be simply predicated of the object a with the same effect as the original sentence. Where the original sentence is thought of schematically as 'Fa' the relative clause is the explicit segregation of the 'F'. The 'such that' construction is the relative clause simplified in respect of word order and fitted with a bound variable to avert ambiguities of cross-reference.

Other uses of the bound variable are readily represented as parasitic upon this use. The quantifiers are 'there is something x *such that*' and 'everything is (a thing) x *such that*'; the description operator is 'the (thing) x *such that*'; the operator of class abstraction is 'the class of (the things) x *such that*'. Quantification can be thought of as application of a functor '$\exists$' or '$\forall$' to a predicate; and this functor is what carries the pure quantitative import, with no intrusion of variables. Similarly description can be thought of as application of a functor 'ι' to a predicate; and class abstraction can be thought of correspondingly. What brings the variable, if any, is the predicate itself, in case it is a relative clause rather than a simple adjective or perhaps some Boolean compound.

Peano saw this, but then slipped into a confusion. He introduced the inverted epsilon for the words 'such that', or for the equivalent words in his three romance languages, and he introduced the functors '∃' and '℩' just as I have described them. But he introduced no functor for class abstraction. He saw his inverted epsilon as already class abstraction; here was his confusion. He did not distinguish between the general term, or predicate, and the class name, a singular term.

The same conflation may be seen in Peano's upright epsilon. For the epsilon that is now standard in set theory comes from Peano; and he adopted it as his copula of predication, the initial of the Greek ἐστί. He inverted it for his 'such that' because this is the inverse of predication; the two cancel. Peano was thus sensitive to the relation between predication and relative clauses: '*a* is a thing *x* such that *Fx*' reduces to '*Fa*'. He was indeed sensitive to the role of the variable as relative pronoun; he was explicit on this. But he must be convicted of the conflation, on two counts. He provides explicitly that his 'such that' expressions designate classes; '($x \ni p$) $\in$ Cls'. And, what is more to the point, he quantifies over them with bound variables; whereas a relative clause or 'such that' clause properly conceived is rather a predicate, susceptible at best of substitution for an unbindable schematic predicate letter. (In this historical context we may pass over the further alternative of a bindable substitutional predicate variable.) With all his sensitivity to grammar he was insensitive to the distinction between general and abstract singular; insensitive to the ontological import of values of variables.

Happily Peano's inverted epsilon has caught on, in informal mathematical contexts, as a sign purely for 'such that'; and I shall so use it.

Russell carried Peano's 'such that' over into *Principles of Mathematics* without improving matters. The so-called propositional function that ended up in *Principia Mathematica* is largely more of the same, but muddied now by triple confusion: property, open sentence, predicate. The term 'propositional function' was adapted from Frege, whose functions might indeed be seen as fictitious designata of relative or 'such that' clauses. I say 'fictitious' because a general term does not designate; and this accords with Frege's characterization of his so-called functions as *ungesättigt*, meaning in a way that there were not really any such things.

One senses in the modern history of logic a distaste for the general term, or predicate. It is partly the effect and partly the cause, I expect, of our slowness to appreciate the schematic status of predicate letters. One tends to conflate such letters with objectual variables, and so reconstrue general terms as abstract singular terms designating properties or classes. Thereby the relative clause, or 'such that' clause, becomes class abstraction; and here we have Peano. Or, fleeing this abstract ontology, one presses the bound variables into purely quantificational duty and thus can operate in sentences. Predicates have been an uneasy intermediate between abstract singular terms on the one hand and out-and-out sentences on the other.

We have to appreciate and exploit schematic letters in order to isolate and appreciate the work of the bound variable itself. It is in representing the 'such that' clauses by steadfastly schematic predicate letters, and not letting these letters into quantifiers, that we can witness the work of the bound variable as a relative pronoun unencumbered by class abstraction.

When logicians became interested in distinguishing between elementary or first-order theories and higher-order theories, schematic use of predicate letters became virtually indispensible. It had become common among Continental logicians by 1930, though generally subject to no clear appreciation of its semantic status. The predicate letter tended to be seen rather as a free variable for properties or classes, bindable in wider contexts and free merely throughout the subcalculus under investigation. Still it is as a schematic predicate letter that we may most clearly view it in retrospect.

But it did not have the effect of freeing logicians to introduce 'such that' clauses into their first-order theories as ontologically innocent predicates; they avoided them as if they were class names. They treated the schematic letter '*F*' as subject to substitution only indirectly, therefore, through substitution of whole open sentences for '*Fx*' or '*Fxy*'. The rules for such substitution are complex, for they must coordinate the sentences substituted for '*Fx*' and '*Fz*', or for '*Fxy*' and '*Fzw*'. When such rules were at length devised, they invoked auxiliary open sentences that served much the purpose of 'such that' clauses after all, though behind the scenes. Hilbert and Bernays, 1934, called them *Nennformen*. Two sentences may be substituted respectively for '*Fxy*' and '*Fzw*' if they can both be got from some *Nennform* by substituting '*x*' and '*y*' on the one hand and '*z*' and '*w*' on the other. By coincidence I proceeded similarly in the same year, 1934, in my first book. In later books I made the *Nennformen* more graphic by use of circled numerals. I called these formulas *stencils* in *Elementary Logic*, 1941. They were sentences with indexed blanks. Expressions to much the same effect had been called *rhemes* by C.S. Peirce in 1892, though not in explicit connection with a problem of substitution. Also they could vaguely be called propositional functions, in one sense of that resilient term. It was not until my 1945 paper that it dawned on me to call them predicates, thus recognizing at last that they were playing the role of the relative clause, the 'such that' clause. And still I kept them behind the scenes. They were devices for calculating substitution, and formed no part of the resulting sentences or schemata.

But in fact they can contribute much to the explicit formalism, quite apart from substitution. They can contribute even to set theory, where genuine class abstraction is already available too. Thus consider the axiom schemata of *replacement* in the Zermelo-Fraenkel system.

$$(u)(x)(y)(z)(Fxz \;.\; Fyz \;.\supset.\; x = y) \quad (\exists w)(x)(x \in w \;.\equiv\; (\exists z)(Fxz \;.\; z \in u))] \;.$$

It remains thus turgid as long as we confine predicate letters thus to positions of predication -- 'Fxz', Fyz'. The effect of 'such that', however, is to free the complex predicate from such embedding. We can then define and apply certain functors to complex predicates as follows, paralleling definitions that are already familiar for classes.

$$F``u =_{df} x \ni (\exists z)(Fxz \;.\; z \in u) \;.$$

$$\text{Func } F =_{df} (x)(y)(z)(Fxz \;.\; Fyz \;.\supset.\; x = y).$$

The axiom schema of replacement is then easily grasped:

$$(u)[\text{Func } F \supset (\exists w)(x)(x \in w \;.\equiv\; (F``u)x)] \;.$$

Under the conventions of my *Set Theory and Its Logic* it boils down yet further, but I shall not pause over them.

The notation of predicates and predicate functors serves some purposes of the class name while not requiring the class to exist. Such was the above case; for the purpose of the axiom schema would be defeated by assuming existence of the class $\{xy\colon\ Fxy\}$.

Gödel in his 1940 monograph availed himself of this convenience, though not calling them relative clauses or general terms. He introduced the expressions as eliminable quasi-names of imaginary classes which he called notions. At a more elementary level but in a more explicit way I did much the same under the head of virtual classes in my Portuguese lectures of 1942, which came out in 1944. In Set Theory and Its Logic twenty years later I worked up a streamlined general formalism for virtual classes, closely integrated with the formalism for real classes. There is a note of pathos in presenting the notation for notions or virtual classes thus as a simulation of class names when it is really a matter of relative clauses, general terms, and should be seen as prior to any thought of classes. Still it should be said that in Set Theory and Its Logic some simplicity is gained from the close integration of virtual and real classes.

Care must be taken not to confuse notions or virtual classes with ultimate or so-called "proper" classes. These latter are real classes, values of bound variables, and differ from sets only in not being members of further classes. The notions or virtual classes, on the other hand, are only a manner of speaking and not really there at all, not being values of bound variables. Their seeming names are really predicates, and their seeming variables are schematic predicate letters. What are represented as ultimate or "proper" classes in Bernays's system of 1958 are mere notions or virtual classes, for they are not values of bound variables; but the ultimate or "proper" classes in his system of 1937-54 are the real thing. It is an open question whether Cantor and König were anticipating the one or the other in some brief passages around the turn of the century.[3]

We are noting how unready logicians have been to think directly in terms of a calculus of complex predicates. They gravitate toward sentences on the one hand or toward class names on the other, even to the point of simulating names of imaginary classes. This bias has long been visible at the most elementary level, indeed, in the attitude toward the perversely so-called Boolean calculus of classes. There is no call for classes there. This bit of logic has its whole utility as an algebra of predicates, represented by schematic letters subject to functors of union, intersection, complement, inclusion, coextensiveness, and Peano's '∃' of non-emptiness and its dual '∀'. Three of these functors form predicates from predicates, and four form sentences from predicates. Then we add truth functions for compounding these sentences into further ones.

This Boolean calculus of predicates and predicate functors is the

[3]See Set Theory and Its Logic, p. 212n.

easy version of monadic quantification theory; and there are no quantifiers and no variables, but only the schematic letters. It is only in the new third edition of Methods of Logic that I have switched to this calculus for the basic presentation of monadic logic. I avoid the identity notation '$F = G$' in order not to encourage at this level any thought of classes; for 'F' and 'G' are still the schematic predicate letters of the logic of quantification. The convenience of Boolean notation has often been foregone because of the mistaken belief that it calls for classes. And conversely there is the new irony in elementary schools of a so-called set theory that is just this Boolean bit of monadic logic and should not be seen as set theory at all.

Behmann should be mentioned as one who, in 1927, treated the Boolean functors in very much the style of predicate functors. However, he then promptly spoiled matters by quantifying and thus effectively according his predicates the status of class names after all. It is the familiar pitfall.

My theme was the variable. We may do well now to retrace it. I am persuaded that the embryo of the bindable variable, psychogenetically, is the relative pronoun.[4] Its status as bindable variable stands forth explicitly when we regiment the relative clause using 'such that'. Relative clauses are adjectives, general terms, that is, predicates. How is a logician to frame a formal calculus of such expressions, a calculus of predicates? He will need to represent the predicates by letters as dummies. He is dim on schematic letters, so he thinks these letters must be variables with some sort of abstract objects as values; and so his relative clauses become abstract singular terms and his 'such that' becomes class abstraction. A later logician, alive to schematic letters, reacts with a true calculus of predicates; and it is the familiar quantification logic in its usual modern style of exposition. But he overreacts: he insists on keeping his predicate letters embedded with their arguments, fearing that a predicate floating free and ungesättigt would be a class name again. He has failed to appreciate 'such that' as an ontologically innocent operator for isolating pure complex predicates, representable by free-floating schematic letters. And so, incidentally, he has needlessly surrendered his little old Boolean logic to his unregenerate predecessor the set theorist.

Its restoration involves curious ironies. Its restoration de-

[4] See The Roots of Reference.

pends, we saw, on a better appreciation of the bindable variable as an appurtenance of the relative clause, not of the class abstract. But the schematic predicate letters then become detachable from their variable arguments, and so the variables themselves disappear.

Bound variables vanish thus from the scene, in our Boolean calculus of predicates, but they lurk in the wings. They figure in the programming, to switch to a computer metaphor. For when we apply this calculus to verbal examples, we shall want usually to interpret 'F', 'G', etc. not just by substitution of pat words or phrases such as 'man' or 'Greek' or 'white whale', but by substitution of such relative clauses as '$x \ni (\exists y)(y$ is son of $x)$' or '$x \ni (3x^2 > 2x)$'; and here we have the bound variable at its proper work. Also we might still use this variable in hidden foundations of our Boolean calculus of predicates, thus:

$$\overline{F} =_{df} x \ni \sim Fx \ ,$$

$$F \cap G =_{df} x \ni (Fx \ . \ Gx) \ ,$$

$$F \cup G =_{df} x \ni (Fx \vee Gx) \ ,$$

$$F \subseteq G =_{df} \forall(x \ni (Fx \supset Gx)) \ .$$

We well know that quantification theory, which is so much more complex than the Boolean predicate calculus, has its serious motivation in polyadic predicates. When we move to polyadic logic, the bound variable quits the wings and gets into the act. The basic job of the bound variable is cross-reference to various places in a sentence where objective reference occurs; and whereas monadic logic calls for this service only in the preparations, polyadic logic calls for it also within the ongoing algorithm, in order to keep track of permutations and indentifications of arguments of polyadic predicates. It is in such permutations and identifications that the bound variable enters essentially into the algorithm, and here it is, by the way, that decision procedures cease to be generally available.

There is evidence of a connection. Polyadic logic remains decidable, like monadic logic, as long as there is no crossing up of argument places. There is a decision procedure covering every quantificational schema that is fluted, as we might say, in the following sense. Every predicate letter has the same variable 'x' as its first argument, though this repeated letter may in its different occurrences be bound by different occurrences of '(x)' or '$(\exists x)$'. Every predicate letter has one and the same letter 'y' as its second argument, if any; and so on. And, a final requirement, each occurrence of a 'y' quantifier stands in the scope of some 'x' quantifier; each occurrence of a 'z' quantifier stands in the scope of some 'y' quantifier; and so on. I gave a decision procedure for such formulas at the Congress of Vienna. (A further proviso was that all the predicate letters have the same number of argument places; but this appears superfluous.)

The variable, then, it seems, is the focus of indecision. It does not, however, set bounds to algebrization. By supplementing the Boolean predicate functors with a few more predicate functors we can, if we wish, still banish the bound variable for good. For there are predicate functors that will do all necessary linking and permuting of argument places.

The predicate functors that I have in mind are somewhat reminiscent of Schönfinkel's combinators, but with a deep difference: his method had the full strength of set theory, whereas what I have in mind is equivalent to quantification theory, or first-order predicate logic. Mine is closer to Tarski's cylindrical algebra, especially as modified by Bernays, and to work of Nolin. There are the Boolean functors as before; but complement and intersection suffice on that score. I construe them in application to m-place and n-place predicates 'F^m' and 'G^n' generally, as follows.

$$\overline{F^m}\, x_1 \ldots x_m \equiv {\sim}F^m\, x_1 \ldots x_m \quad .$$

$$(F^m \cap G^n)x_1 \ldots x_{\max(m,n)} \equiv .\ F^m\, x_1 \ldots x_m \,.\, G^n\, x_1 \ldots x_n \quad .$$

The variables 'x_1' etc. have no place in the system, but serve only in my present explanation of the functors. To continue, then: the further devices, which together accomplish all work of bindable variables, are a <u>cropping</u> functor, a <u>padding</u> functor, a <u>permutation</u> functor, and a constant <u>identity</u> predicate. They are explained as follows:

$$(\daleth F^m)x_2 \ldots x_m \equiv (\exists x_1)F^m\, x_1 \ldots x_m \; ,$$

$$(\lessdot F^m)x_0 \ldots x_m \equiv F^m\, x_1 \ldots x_m \; ,$$

$$(PF^m)x_1 x_3 \ldots x_m x_2 \equiv F^m\, x_1 \ldots x_m \; ,$$

$$Ixy \equiv .\ x = y .$$

The notation of this version of quantification theory consists solely of schematic predicate letters and the predicate 'I' and the five functors. The schematic letters carry exponents to indicate the number of argument places, since those places are never visibly occupied.

Truth functions come through as the cases of complement and intersection where $\underline{m} = \underline{n} = 0$. I used to have two permutation functors, but George Myro got them down to one.[5]

The Boolean calculus of predicates is decidable; so it is '$\underline{I}$' and the last three functors that bring the full force of the bound variable, transcending any general decision procedures. It would be interesting to know whether the decision procedure for the Boolean part can be extended to cover one or two also of these three further functors, thus further isolating the source of the undecidable.

Havard University

[5]The system is presented in "Algebraic logic and predicate functors," pp. 8-25. Myro's improvement appears in footnote 5 of my "Truth and disquotation."

REFERENCES

Behmann, Heinrich, Mathematik und Logik, Leipzig, 1927.

Bernays, Paul, "A system of axiomatic set theory," Journal of Symbolic Logic 2, 6-8, 13, 19 (1937, 1941-43, 1948, 1954).

_________, "Ueber eine natürliche Erweiterung des Relationenkalküls," in A. Heyting, ed., Constructivity in Mathematics (Amsterdam: North-Holland, 1959), pp.1-14.

_________, and A.A. Fraenkel, Axiomatic Set Theory, Amsterdam: North-Holland, 1958.

Church, Alonzo, The Calculi of Lambda Conversion, Princeton, 1941.

Frege, Gottlob, Grundgesetze der Arithmetik, Vol.1, Jena, 1893.

Gödel, Kurt, The Consistency of the Continuum Hypothesis, Princeton, 1940.

Hilbert, David and Paul Bernays, Grundlagen der Mathematik, Vol.1, Berlin, 1934.

Marcus, Ruth B., "Modalities and intensional languages," Synthese 13 (1961), pp.303-322.

Nolin, L., "Sur l'algèbre des prédicats," in the colloquium volume Le Raisonnement en Mathématiques et en Sciences Expérimentales, Paris: Centre National de la Recherche Scientifique, 1958, pp.33-37.

Parsons, Charles, "A plea for substitutional quantification," Journal of Philosophy 68 (1971), pp.231-237.

Peano, Giuseppe, Formulaire de Mathématiques, Turin, 1894-99; Paris, 1901.

Peirce, C.S., Collected Papers, Cambridge: Harvard, 1931-35.

Quine, W.V., A System of Logistic, Cambridge: Harvard, 1934.

_________, "Logic based on inclusion and abstraction," Journal of Symbolic Logic 2 (1937), pp.145-152. Reprinted in Selected Logic Papers.

_________, Mathematical Logic, New York, 1940.

_________, Elementary Logic, Boston, 1941.

_________, O Sentido da Nova Logica, Sao Paulo, 1944.

_________, "On the logic of quantification," Journal of Symbolic Logic 10 (1945), pp.1-12. Reprinted in Selected Logic Papers.

_________, Methods of Logic, New York: Holt, 1950, 1959, 1972.

_________, Set Theory and Its Logic, Cambridge: Harvard, 1963, 1969.

_________, Selected Logic Papers, New York: Random House, 1966.

________, "On the limits of decision," Akten des XIV. Internationalen Kongresses für Philosophie, Wien, 1968, Vol.3, pp.57-62.

________, "Algebraic logic and predicate functors," Indianapolis: Bobbs-Merrill, 1971 (25pp.).

________, "Truth and disquotation," in L. Henkin, ed., Tarski Symposium, Providence: American Matehmatical Society, at press.

________, The Roots of Reference, Evanston: Open Court, at press.

Russell, Bertrand, The Principles of Mathematics, Cambridge, England, 1903.

Schönfinkel, Moses, "Ueber die Bausteine der mathematischen Logik," Mathematische Annalen 92 (1924), pp.305-316.

Tarski, Alfred and F.B. Thompson, abstracts in Bulletin of American Mathematical Society 58 (1952), pp. 65f.

Whitehead, A.N. and Bertrand Russell, Principia Mathematica, Vol.1, Cambridge, England, 1910.

ABOLITION OF THE FREGEAN AXIOM

Roman Suszko

This paper is partly stimulated by a talk given by Dana Scott on Lewis' systems in the Symposium on Entailment, December 1971, [1]. Any endeavour, however, to reconstruct Lewis' program or to defend it is far beyond my intention. What matters here is the following. Scott makes a great deal of propaganda on behalf of (a) the general theory of entailment relations (or consequence operations) and (b) truth-valuations. Furthermore, "a nagging doubt" in Scott's mind, concerned with possible-world semantics induces him to use both (a) and (b) and a trick of making inferences visible, to arrive eventually at the strong modal systems, S_4 and S_5.

There are, of course, plenty of ways to obtain modal systems. Here, I want to call your attention in particular to the somewhat disquieting fact that the strong modal systems (but by no means all modal systems) are theories based on an extensional and logically two-valued logic, labelled NFL, exactly in the same sense that axiomatic arithmetic is said to be based on (pure!) logic [created essentially by Frege, (hence labelled FL) and well-known from text-books of mathematical logic]. This paper is not, however, another exercise in so-called modal logic. I essentially agree by the way, with Quine's comments [2] on that kind of logic. The main subject here is the construction of NFL and its basic properties. Also, the relation between NFL and FL will be discussed. The general theory of entailment will serve as a framework for three methods of building NFL. In fact, we will arrive at NFL using truth-valuations, models and logical axioms and rules of inference.

As an intelligent reader you instantly conjecture that there must be some hocus-pocus underlying NFL. Indeed, there is. It consists essentially in following Frege in building pure logic but only to certain decisive point. Of course, you need not use his archaic notation or terminology. Also, you may easily avoid his syntactic shortcomings. For example, you are naturally inclined to keep formulas (sentences) and terms (names) as disjoint syntactic categories. But, when you come to his assumption, called here the <u>Fregean axiom</u>, that all true (and, similarly, all false) sentences describe the same thing, that is, have a common referent, just forget it, please; at least until NFL is constructed. At that time, I am sure, you will better understand what the Fregean axiom is and you'll be free to accept it, if you still like it so much.

The trick underlying NFL is fairly easy and also quite innocent. It is true that it seduced me succesfully and I am now addicted to it. I even reject the Fregean axiom. However, I do not insist that you go so far. But try NFL cautiously. I assure you that NFL offers you an intellectual experience, unexpected in its simplicity and beauty, far surpassing all "impossible worlds". But I am frank and fair, by my nature. So I tell you keep the Fregean axiom hidden in your pocket when entering the gate of NFL and be ready to use it at once, when you feel a confusing headache. Formally, you will be collapsing NFL into FL. Informally, you will be expelling yourself from a logical paradise into the rough, necessary world.

Surprisingly enough, logicians do not <u>want</u> NFL. I know it from five years of experience and this is the right way of putting it, believe me. Being even so close to NFL sometimes, Logicians stubbornly strive after something else. When not satisfied with FL they choose to work with the powerset 2^I as exposed convincingly by Scott [3]. They even can, I admit, work on it as hard as in a sweatshop. So mighty is, Gottlöb, the magic of your axiom! Whatever (cheatingly) one calls elements of the index set I, the powerset 2^I remains a distinct shadow of the Fregean axiom.

If we want to follow Frege we must consent to his basic ideals of unabiguity and extensionality. To stress this point we start with his famous semantical scheme of Sinn und Bedeutung. It is obvious today that the abyss of thinking in a natural language does not fit into the Fregean scheme. But this is another story. Here, it must suffice to notice that we all live (and cannot completely get out of) that messy abyss with all its diffuse ghosts (in Hermann Weyl's [4] phraseology) of ambiguity, vague flexibility, intensionality and modality. We really enjoy them. But not always. When forced to construct a theory, we try to make our ideas precise and climb to the heights of extensionality. Then, the structure of our theoretical thought corresponds sufficiently well to the syntax of the Begriffschrift, i.e., a formalized language which does fit into the Fregean semantical scheme.

1. Reference, Sense and Logical Values.

We assume the principle of logical two-valuedness and choose 1 and 0 to represent the truth and falsity of sentences, respectively. Then, the Fregean scheme can be presented by the following diagram:

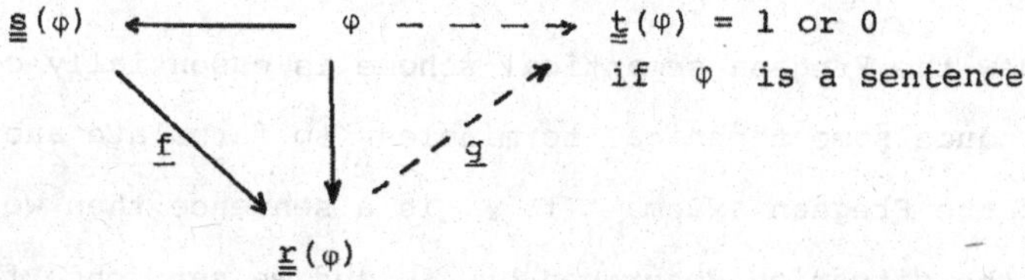

Here, φ is either a name (term) or sentence, $\underline{\underline{r}}(\varphi)$ = the referent of φ, i.e., what is given by φ, $\underline{\underline{s}}(\varphi)$ = the sense of φ, i.e., the way $\underline{\underline{r}}(\varphi)$ is given by φ. Moreover, if φ is a sentence then $\underline{\underline{t}}(\varphi)$ = the logical (or truth) value of φ. The assignments $\underline{\underline{r}}$, $\underline{\underline{s}}$ and $\underline{\underline{t}}$ are related as follows:

(1.1) $$\mathbf{s}(\varphi) \neq \mathbf{s}(\psi) \quad \text{whenever} \quad \mathbf{r}(\varphi) \neq \mathbf{r}(\psi)$$

and, for sentences φ, ψ:

(1.2) $$\mathbf{r}(\varphi) \neq \mathbf{r}(\psi) \quad \text{whenever} \quad \mathbf{t}(\varphi) \neq \mathbf{t}(\psi)$$

Note that the converse of (1.2) is a version of the Fregean axiom. On the other hand, the converse of (1.1) is forbidden. One may say that we just need the assignment $\mathbf{s}$ to cope with the existence of expressions φ, ψ which differ somehow in meaning but refer to the same, i.e., $\mathbf{r}(\varphi) = \mathbf{r}(\psi)$. Then, we simply set: $\mathbf{s}(\varphi) = \mathbf{s}(\psi)$.

Frege put much stress on the assumption, restated by Church ([5], p. 9) that referents depend functionally on senses. That is, there is a function $\underline{f}$ such that for any name or sentence, φ:

(1.3) $$f(\mathbf{s}(\varphi)) = \mathbf{r}(\varphi)$$

This is, however, easily seen to be virtually equivalent to (1.1). Similarly, (1.2) is equivalent to the existence of a function $\underline{g}$ such that for any sentence :

(1.4) $$\underline{g}(r(\varphi)) = \mathbf{t}(\varphi).$$

Now the Fregean semantical scheme is essentially complete and we introduce some technical terminology to formulate another equivalent of the Fregean axiom. If φ is a sentence then we say that $\mathbf{r}(\varphi)$ = the *situation described by* φ and we set, once for all, $\mathbf{s}(\varphi)$= the *proposition expressed by* φ. Similarly, $\mathbf{r}(\varphi)$ = the *object denoted by* φ, if φ is a name. Assignments $\mathbf{r}$ and $\mathbf{s}$ apply to names as well as sentences.[I] Therefore, one may expect a strong analogy between the pairs: (a) names (terms) and sentences (formulas) and, (b) situations and objects. However, because of the assignment $\underline{\underline{t}}$ one may also expect some essential differences in (a) or (b). Certain differences in (a), e.g., those concerned with assertion, inferences and entailment are quite familiar. The listed notions are concerned with sentences and do not apply meaningfully to names. Thus, we are prepared to observe an interesting interplay of analogy and difference within (a) and (b). Now, it is easy to see that the Fregean axiom has the following equivalent: there are no more than two situations described by sentences. Hence, we baldly say that we do not accept the Fregean axiom since we do not want to impose any quantitative limitation from above on either situations or objects. Thus, we have a analogy between situations and objects. A difference exists also. Under rather obvious assumptions on the language we infer that (1) there exists at least one object and, by (1.2) or (1.4), (2) there exist at least two distinct situations or, more explicitely, the totality of all situations is divided into two non-empty collections.[(II)]

The Fregean semantical scheme is a simplified and condensed summary of the semantics of the Begriffschrift (formalized language) and its full value can be revealed only if combined with the syntax of the expressions involved. Given a language, the assignments $\underline{\underline{t}}$ and $\underline{\underline{r}}$ can be elaborated with the machinery of truth-valuations and models, respectively, and a two-valued extensional logic may be imposed on the language. Thereafter, the assignment $\underline{\underline{s}}$ can be elaborated and effectively used in studying the logic constructed.[(III)]

If one accpets the Fregean axiom and follows Frege in constructing pure logic then one will arrive at FL, the Fregean logic. We will continue Frege's program <u>without</u> his axiom. It is like realising Euclid's program without his fifth postulate. In that case, one arrives at so called absolute geometry and there are just two possibilities: the way of Euclid or that of Lobachevski and Bolyai. Here, we get NFL, that is, the absolute non-Fregean logic. However, we are opening a Pandora's box, since the realm of possibilities we face is uncountable (as will be shown). NFL is so weak that the totality of logics at least as strong as NFL appears, at first glance, chaotic. We may hope it is not since the crystal of NFL lies at the bottom. Certainly, we have an embarassement of riches. This totality includes FL. This fact is very important since we know FL so well. Therefore, FL may serve as a guide and help us get some insight into the chaos we have so imprudently created. In fact, FL helped me to find there certain precious items, other than NFL and FL.

We are now ready to construct NFL. There is nothing except our will (or perhaps, some cataclysm) which could prevent us. I mean that we did not leave open any essential problem. In particular, we need not ask our intuition how to proceed in construcing NFL because we will simply follow Frege and use the standard formal tools.

2. Entailment, Identity.

A logic is meant here not a set of formulas (logical theorems, tautologies), generated via certain rules from some set of axioms but as an entailment relation (or "consequence operation" in A. Tarski's sense, 1930), operating in the set of formulas (sentences) of some language. This is a common point with Scott. Any entailment is a relation $\vdash$ between sets of formulas $X, Y, Z, \ldots$ and formulas $\alpha, \beta, \gamma, \ldots,$ subject to three basic laws of Reflexivity, Monotonicity and Transitivity, as in []], p. 796. If we say

$$X \text{ entails } \alpha \text{ or not, i.e., } X \vdash \alpha \text{ or non } X \vdash \alpha$$

we allow the antecedent X and the succedent α to be correspondingly, an arbitrary set of formulas and any single formula.(IV) If

X is finite and, $X = \{\alpha_1,\ldots,\alpha_n\}$ or $X = \emptyset$ = the empty set then instead of $X \vdash \alpha$ we may write $\alpha_1,\ldots,\alpha_n \vdash \alpha$ and $\vdash \alpha$, respectively. If $\vdash \alpha$ then α is said to be a tautology of $\vdash$. The set of all $\vdash$ tautologies is an example of a theory of $\vdash$. A set X of formulas is said to be a theory of $\vdash$ if X is closed under $\vdash$, that is, X contains every formula entailed by X. There are many theories of the given entailment, in general. Since the intersection of any collection of theories is a theory again there always exists a least (smallest) theory: the set of all tautologies.(V)

We are interested in a specific entailment relation, that is, NFL entailment, imposed on a formalized language. First, we define a class of formalized languages which I prefer to call "languages of kind W". Vocabularies of languages of kind W involve: (1) sentential variables, (2) nominal variables, (3) formators, not binding variables, of different sorts, like functors, predicates and connectives, e.g., $\neg$, $\wedge$, $\vee$, $\Longrightarrow$, $\Longleftrightarrow$ of negation, conjunction, disjunction, implication and equivalence, respectively), (formators binding variables (sentential and/or nominal), i.e., quantifiers (universal $\forall$, esistential $\exists$ and even numerical quatifiers) and the unifier (description operator). Starting with a given vocabulary and using familiar rules of formation, one may build terms, including nominal variables and names, and formulas, including sentential variables and sentences.(IV) The cornerstone of W-languages is the notion of identity which occupied Frege's mind so much. The identity formator falls into the group (3) above as a binary formator which must, however, split into an identity <u>predicate</u> and an identity <u>connective</u>, for obvious syntactical reasons. To emphasize the analogy, both versions of identity will be rendered by the same symbol $\equiv$. Thus, whenever both φ,ψ are terms or formulas then $\varphi \equiv \psi$ is a formula, called an <u>equation</u>. The symbol $\equiv$ is then an identity predicate or identity connective, respectively.(VII)

The NFL entailment relation in comprehensive W-languages has been defined in a syntactic way, by logical axioms and rules of inference in [9] and [10] and, a conventional completeness theorem for it has been proved by S. L. Bloom in [11]. NFL involves precisely the logical laws for identity: trivial equations $\varphi \equiv \varphi$ are tautologies and the law of replacement of equals by equals salva identitate is guaranteed by NFL entailment.

If you do not have a clear idea of identity (equality, sameness, one-ness) then you had better stop reading right now. As an explanation of this notion

$$(2.1) \qquad \underline{\underline{t}}(\varphi \equiv \psi) = 1 \text{ iff } \underline{\underline{r}}(\varphi) = \underline{\underline{r}}(\psi)$$

is essentially correct but somewhat circular and, hence, not very helpful. We must begin with the unquestionable logic of identity along Leibniz' ideas. It is, as Quine [12] made clear beyond any doubt, the only measure of extensionality (referential transparency) of notions and contexts. A context is intensional (non-extensional, referentially opaque) just if it violates some logical law of identity. Since NFL entailment involves the logic of identity governing all terms and formulas of the corresponding language, we may say that NFL does not tolerate any intesional context or notion. All notions encoded in a W-language supplied with NFL entailment must be extensional. In particular, equations $\varphi \equiv \psi$ of both kinds are extensional. In sum, NFL is the paradigm of an extensional (and two-valued) logic.

In this article, we will not deal with the comprehensive languages of kind W. More modest, open W-languages without bound variables and with correspondingly restricted NFL entailment, as in [9], appear sufficiently interesting. They contain equations of both kinds and, hence, on that syntactic level one may study the interplay of analogy and difference noted previously; compare [13]. But even open W-languages seem too complicated for this article. The identity predicate is more than well known and quite innocent. It is the identity connective which carries a devil's brand. So, let us abandon the identity predicate and make one more step down to reach the level of the sentential calculus with identity connective, the so called SCI, which is the crucial place of the battle for NFL.

SCI is the NFL entailment relation imposed on formulas of the sentential language with the volabulary consisting of sentential variables $p, q, r, \ldots$ and, connectives $\neg, \wedge, \vee, \Rightarrow, \Leftrightarrow$ and $\equiv$. SCI has been extensively studied in [14] and [15]. A short list of properties of SCI may be found in [16] and [17]. Truth-functional connectives seem to be childish toys in comparison with the identity connective. As far as this toy is concerned one has to be warned not to commit the childish mistake of confusing formulas of the form $\alpha \Rightarrow \beta$ with the meta-statements saying that a formula α entails the formula β, i.e., $\alpha \vdash \beta$. On the other hand, it is no discovery to anyone to note that the nature of the identity connective $\equiv$ is not truth-functional. One may ask how it could be otherwise? You rightly answer: $\equiv$ is truth-functional and, then, coincides with material equivalence, if and only if the Fregean axiom holds. In the case of NFL the only relation between truth-values of α, β and $\alpha \equiv \beta$ is given by (2.1) together with (1.2). Thus, we only exclude that simultaneously

$$\underline{\underline{t}}(\alpha \equiv \beta) = 1 \quad \text{and} \quad \underline{\underline{t}}(\alpha) \neq \underline{\underline{t}}(\beta).$$

Although $\equiv$ is a non-truth-functional connective the machinery of truth-valuations, explained by Scott, works as well with $\neg, \wedge, \vee, \Rightarrow, \Leftrightarrow$ as with $\equiv$. Clearly, we do not need all functions $\underline{t}: FM \to \{0,1\}$ from the set FM of all SCI-formulas to the set $\{0,1\}$ of truth values. We will only use certain truth-value assignments to formulas. These may be called shortly truth-valuations (for SCI). They characterize $\neg$, $\wedge$, $\vee$, $\Rightarrow$, $\Leftrightarrow$ adequately as the intended truth-functional connectives and $\equiv$ as the identity connective.

3. Truth-valuations.

A function $\underline{t}$:FM $\to$ {0,1} is said to be a truth-vaulation if for all $\alpha,\beta,\gamma,\delta$ in FM:

$$\underline{t}(\neg\alpha) \neq \underline{t}(\alpha) \tag{3.1}$$

$$\underline{t}(\alpha \wedge \beta) = 1 \text{ iff } \underline{t}(\alpha) = \underline{t}(\beta) = 1 \tag{3.2}$$

and, similarly, for $\vee,\Rightarrow,\Leftrightarrow$ as required by the usual truth tables and,

$$\underline{t}(\alpha \equiv \alpha) = 1 \tag{3.3}$$

$$\underline{t}(\neg\alpha \equiv \neg\beta) = 1 \text{ if } \underline{t}(\alpha \equiv \beta) = 1 \tag{3.4}$$

$$\underline{t}([\alpha \mathbin{\&} \gamma] \equiv [\beta \mathbin{\&} \delta]) = 1 \text{ if } \underline{t}(\alpha \equiv \beta) = \underline{t}(\gamma \equiv \delta) = 1 \tag{3.5}$$

where $\&$ stands for one of binary connectives $\wedge$ $\vee$ $\Rightarrow$ $\Leftrightarrow$ and $\equiv$ and, finally

$$\underline{t}(\alpha \equiv \beta) = 0 \text{ whenever } \underline{t}(\alpha) \neq \underline{t}(\beta). \tag{3.6}$$

Given a truth-valuation $\underline{t}$, we immediately have an entailment $\vdash_{\underline{t}}$, depending on t: $X \vdash_{\underline{t}} \alpha$ iff $\underline{t}(\alpha) = 1$ whenever $\underline{t}(\beta) = 1$ for every β in X. But, we are interested in an entailment which does not depend on any particular truth-valuation. So we set:

$$X \vdash^{(t)} \alpha \text{ iff } X \vdash_{\underline{t}} \alpha \text{ for every truth-valuation } \underline{t}. \tag{3.7}$$

In this section, we simply write $\vdash$ instead of $\vdash^{(\underline{t})}$ for the SCI-entailment, defined by truth-valuations. In (3.7), X is an arbitrary, possibly infinite set of formulas. However, $\vdash$ has the Finiteness property:

(3.8) if $X \vdash \alpha$ then there is a finite subset Y of X such that $Y \vdash \alpha$.(VIII)

Obviously: $\alpha, \alpha \Rightarrow \beta \vdash \beta$. Moreover, $\vdash$ has the Deduction property:

(3.9) if $\alpha_o, \alpha_1, \ldots, \alpha_n \vdash \beta$ then $\alpha_1, \ldots, \alpha_n \vdash (\alpha_o \Rightarrow \beta)$.

Properties (3.8) and (3.9) amalgamate into the following:

(3.10) $X \vdash \alpha$ iff either $\vdash \alpha$ or there exist $\alpha_1, \ldots, \alpha_n$ in X s.th. $\vdash ((\alpha_1 \wedge \ldots \wedge \alpha_n) \Rightarrow \alpha)$.

Thus, the entailment $\vdash$ reduces, in a sense, to the set of its tautologies, TAUT. Hence, we state certain properties of this set.

Call a set of formulas <u>invariant</u> if it is closed under the substitution rule. In other words, invariance of a set X of formulas means the following: for any α in X, the result $\alpha[^v/\beta]$ of the uniform replacement of any variable v in α by any formula β is also in X. As expected, the set TAUT is invariant. Another property of TAUT is its decidability. Indeed the truth-valuations provide an effective method for deciding of any given formula α whether α is in TAUT or not. Note that the reason for this is simply the obvious finiteness of formulas.

Functions $\underline{t}$:FM $\rightarrow \{0,1\}$ with only the properties (3.1) and (3.2) are called <u>general truth-valuations</u>. They respect all truth-functional connectives but do not respect the identity connective. Formulas α such that $\underline{t}(\alpha) = 1$ for each general truth-valuation $\underline{t}$ may be collected into the set TFT, the set of all truth-functional tautologies It also is an invariant set and contains for example $p \vee \neg p$, $p \Rightarrow p$ and $\neg(p \wedge \neg p)$. Obviously, TFT is a proper subset of TAUT.

You may be eager to know what are the formulas in TAUT which are not in TFT. Here are some examples: $p \equiv p$, $(p \equiv q) \wedge (q \equiv r) \Rightarrow (p \equiv r)$ $(p \equiv q) \Rightarrow (q \equiv p)$, $(p \equiv q) \Rightarrow (\neg p \equiv \neg q)$. They are pure trivialities. Another point is interesting and important. Clearly, no formula in TFT can be an equation $\alpha \equiv \beta$ or inequality $\alpha \not\equiv \beta$ (that is, $\neg(\alpha \equiv \beta)$). But how does TAUT behave in this respect? Recall the very popular view that logical theorems (tautologies) are "cognitively empty" and restate it as the natural postulate that "logic cannot teach us any non-trivial equation." First, the natural postulate is satisfied here. Indeed

(3.11) every equation in TAUT is trivial (i.e. $\alpha \equiv \alpha$).

This is an essential property of NFL. One can extend (3.11) to the level of comprehensive W-languages with the identity connective and identity predicate as well. Then the natural postulate holds uniformly for equations of both kinds.[(IX)] Again, we may see that as an analogy. On the other hand, a difference also appears. It is concerned, however, with inequalities $\varphi \not\equiv \psi$. NFL does not offer any tautological inequality involving the identity predicate. But, already on the level of SCI the identity connective behaves differently. In fact, $p \not\equiv \neg p$ is in TAUT.

A subtle circumstance follows from (3.11). The identity connective is tautologically symmetric, that is, $(p \equiv q) \Leftrightarrow (q \equiv p)$ is in TAUT but not in TFT. However, $(p \equiv q) \equiv (q \equiv p)$ is not in TAUT. The identity connective is <u>not</u> tautologically commutative. There are plenty of examples of this phenomenon. Equivalencies $(\neg\neg p) \Leftrightarrow p$, $(p \wedge p) \Leftrightarrow p$, $(p \vee q) \Leftrightarrow (\neg p \Rightarrow q)$, $\neg(p \wedge q) \Leftrightarrow (\neg p \vee \neg q)$ are in TAUT, since they already are in TFT. But, by (3.11) the corresponding equations $(\neg\neg p) \equiv p$, $(p \wedge p) \equiv p$, $(p \vee q) \equiv (\neg p \Rightarrow q)$, $\neg(p \wedge q) \equiv (\neg p \vee \neg q)$ are not in TAUT. One has to ponder the natural postulate, again.

4. Models.

We are not going to elaborate within the framework of conventional semantics the referential component of the Fregean scheme as applied to SCI. You may or may not like it but conventional semantics of formalized languages is, independent of your favorite philosophy, a firm part of universal algebra with a distinct link to the basic ideas of the theory of categories. All we have to do at the beginning is to guess the right class of algebraic structures for SCI and Stephen L. Bloom actually did it in 1969, [11]. Then everything else is mathematically "business as usual", concerned with the language as a "free" algebraic structure and associated models. It involves relativized notions of satisfaction and truth and, hence, provides us with some insight into what the SCI-language is about. Furthermore, it naturally leads to the entailment relation and elucidates some of its important properties.

A non-empty set A together with an unary operation $-$ and five binary operations $\cap, \cup, \dot{-}, \div, \circ$ is here called an algebra:

$$\underline{\underline{A}} = \langle A, -, \cap, \cup, \dot{-}, \div, \circ \rangle \tag{4.1}$$

The class of all algebras (4.1) is very board. It includes, in particular, familiar Boolean algebras (with an additional binary operation $\circ$) where $-, \cap, \cup$ are the operations of complement, meet and join, correspondingly, and $a \dot{-} b = -a \cup b$ and $a \div b = (a \dot{-} b) \cap (b \dot{-} a)$. Here, $a \cup -a = 1$ = the unit, and $a \cap -a = 0$ = the zero of the Boolean algebra A. Also, we write $a \leq b$ for $a \dot{-} b = 1$.

Given an algebra $\underline{\underline{A}}$, imagine its universe A divided into two non-empty subsets. Denote one of them by F and suppose that F is related to the operations in $\underline{\underline{A}}$ as follows (a,b are elements of A):

(4.2) $-a$ is in F iff a is not in F

(4.3) $a \cap b$ is in F iff both a,b are in F

(4.4) $a \cup b$ is not in F iff neither a nor b is in F

(4.5) $a \dot{-} b$ is not in F iff both a,−b are in F

(4.6) $a \dot{\div} b$ is in F iff both $a \dot{-} b$, $b \dot{-} a$ are in F

(4.7) $a \circ b$ is in F iff $a = b$

Then, call F (neutrally) an "f-set of $\underline{\underline{A}}$". Subsequently, form the pair of $\underline{\underline{A}}$ and F, that is, the structure

$$(4.8) \qquad \underline{\underline{M}} = \langle \underline{\underline{A}}, F \rangle = \langle A, -, \cap, \cup, \dot{-}, \dot{\div}, \circ, F \rangle$$

and call it a <u>model</u>, based on algebra $\underline{\underline{A}}$ with the universe A. In the special case of a Boolean algebra $\underline{\underline{A}}$ the conditions (4.2),..., (4.6) force any f-set of $\underline{\underline{A}}$ to be an ultrafilter of $\underline{\underline{A}}$. But, the condition (4.7) also remains in force. Hence, if $\circ$ is an arbitrary, additional operation in a Boolean algebra $\underline{\underline{A}}$, then $\underline{\underline{A}}$ may not contain any f-set at all. It also may happen, however, that every ultrafilter of a Boolean algebra $\underline{\underline{A}}$, with a circle operation, actually is an f-set of $\underline{\underline{A}}$. This is the case precisely if the circle operation of the Boolean algebra has the following property:

$$(4.9) \qquad a \circ b = 1 \text{ if } a = b \text{ and, } a \circ b = 0 \text{ if } a \neq b.$$

Thus, Boolean algebras with a circle operation such that (4.9) holds are the most natural examples of algebras and will be called Henle algebras.[X] A general lesson follows: not every algebra can appear in models. The existence of f-sets is a specific condition imposed on operations in algebras underlying models. One must never forget it. It appears, in certain cases, to be relatively simple.[XI] Anyway, the first step into the SCI-model theory is done; the class of models is precisely defined.[XII] The most natural examples are just Henle models, that is, models $\langle \mathbf{A}, F \rangle$ such that $\mathbf{A}$ is a Henle algebra and F is any ultra-filter of $\mathbf{A}$. Minimal Henle models contain exactly two elements. Since they all are isomorphic we may speak about the smallest Henle model, where $A = \{0,1\}$ and $F = \{1\}$ and, call it the Fregean model. It is essentially the well-known two element Boolean algebra.[XIII] The important circumstance is, however, that all models are so called logical matrices of special kind and, in general, cannot be reduced to mere algebras, possibly with one distinguished element. [XIV]

Given a model $\mathbf{M}$ as in (4.8), we describe what is called "running of variables over the universe A of $\mathbf{M}$". This is done by using valuations over $\mathbf{M}$, i.e., assignments of elements of A to variables $p,q,r,\ldots$. If $\underline{r}$ is a valuation then $\underline{r}(p), \underline{r}(q), \underline{r}(r),\ldots$ belong to A and are values or referents of variables $p,q,r,\ldots$ under $\underline{r}$. The valuation $\underline{r}$ may be seen as extended over the set of all formulas, that is, as an assignment of elements $\underline{r}(\alpha)$ to every formula α . We simply set: $\underline{r}(\neg\alpha) = -\underline{r}(\alpha)$ and $\underline{r}(\alpha \,\ell\, \beta) = \underline{r}(\alpha) * \underline{r}(\beta)$ where ℓ represents one of binary connectives $\wedge, \vee, \Rightarrow, \Leftrightarrow, \equiv$ and $*$ is, correspondingly, one of binary operations $\cap, \cup, \dot{-}, \div, \circ$. Then, clearly, every formula α has a referent $\underline{r}(\alpha)$ in $\mathbf{M}$ under $\underline{r}$[XV] and we can introduce the basic semantical notion of satisfaction. We say that the valuation $\underline{r}$ satisfies a formula α in $\mathbf{M}$ or, α is satisfied in $\mathbf{M}$ by $\underline{r}$ if $\underline{r}(\alpha)$ falls into the f-set of the model $\mathbf{M}$. Subsequently, other semantical notions can be easily defined.

Given a model $\underline{\underline{M}}$, a formula α is said to be satisfiable or true in $\underline{\underline{M}}$ if correspondingly α is satisfied by some or every valuation over M. We also define an entailment $\vdash_{\underline{\underline{M}}}$, depending on $\underline{\underline{M}}$: $X \vdash_{\underline{\underline{M}}} \alpha$ iff α is satisfied by every valuation (over $\underline{\underline{M}}$) which satisfies all formulas in X.

It is easily seen that the set $TR(\underline{\underline{M}})$ of all formulas true in $\underline{\underline{M}}$ contains precisely all tautologies of $\vdash_{\underline{\underline{M}}}$. However, this is not the main merit of the class of entailment relations, generated by single models. The point is that we may use them all to introduce the entailment $\vdash^{(\underline{\underline{r}})}$, which is referentially defined and is independent of any particular model:

(4.10) $X \vdash^{(\underline{\underline{r}})} \alpha$ iff $X \vdash_{\underline{\underline{M}}} \alpha$, for every model $\underline{\underline{M}}$.

In connection with these notions we want to discuss the role of finite models, i.e., those whose universes are finite. First, we have the finite model property:

(4.11) If a formula is satisfiable in some model then it is satisfiable in some finite model.

It follows that

(4.12) the set of all formulas true in all finite models equals to the set of all tautologies of $\vdash^{(\underline{\underline{r}})}$, i.e., formulas true in all models.

On the other hand, the class of all finite models is insufficient to characterize the entailment $\vdash^{(\underline{\underline{r}})}$. In fact, the restriction of (4.10) to finite models is false.(XVI)

So far, we have defined two entailments $\models^{(t)}$, $\models^{(r)}$ and you are naturally eager to compare them. However, the strategy is here to first define a third entailment and only then discuss how all these three entailments are related.

5. Syntactic entailment.

Entailments are very often defined inductively as generated from some set of axioms via certain rules of inference.(XII) We also use this method. The MP-rule, $\alpha, \ \alpha \Rightarrow \beta / \ \beta$ is the only rule of inference. Select an adequate set TFA(XVIII) of truth-functional axioms and collect into the set IDA of axioms for the identity connective, all formulas of the following form:

(5.1) $\alpha \equiv \alpha$

(5.2) $(\alpha \equiv \beta) \Rightarrow (\neg\alpha \equiv \neg\beta)$

(5.3) $(\alpha \equiv \beta) \wedge (\gamma \equiv \delta) \Rightarrow [(\alpha \& \gamma) \equiv (\beta \& \delta)]$

where $\&$ stands for one of binary TF-connectives $\wedge, \vee, \Rightarrow, \Leftrightarrow$;

(5.4) $(\alpha \equiv \beta) \wedge (\gamma \equiv \delta) \Rightarrow [(\alpha \equiv \gamma) \equiv (\beta \equiv \delta)]$

(5.5) $(\alpha \equiv \beta) \Rightarrow (\alpha \Leftrightarrow \beta)$

Denote the union TFA $\cup$ IDA by LA = the set of logical axioms and define the entailment $\vdash$ as follows:

(5.6) $X \vdash \alpha$ if α is derivable by MP in finite number of steps from $X \cup$ LA.

Since $\vdash$ is defined as derivability relation (by LA and MP) we are, to some extent, told how to use it in the language, that is, how to perform reasonings involving formulas of the SCI-language such that conclusions follow premises. In this we do not feel bound by the particular definition (5.6) of $\vdash$. It is, obviously, extremely important to prove theorems about $\vdash$. The decisive fact is that LA is defined by finite number of schemes of formulas and only a finite number of rules (schemes) of inference are used. However, once $\vdash$ has been defined what essentially matters here is the entailment itself, i.e., the logic and, consequently all its tautologies (besides axioms in LA) and all $\vdash$-valid rules of inference (besides MP), i.e., those in which premisses always $\vdash$-entail conclusions.

It is very easy to show that $\vdash$ has the Finiteness and Deduction property precisely in the sense of (3.8) and (3.9). Therefore,

(5.7) $\alpha_1, \ldots, \alpha_n \vdash \alpha$ iff $\vdash [(\alpha_1 \wedge \ldots \wedge \alpha_n) \Rightarrow \alpha]$

Thus, valid schemes of inference correspond (one-one) to tautological implicational schemes of formulas.[XIX] In particular, all schemes of inference of truthfunctional logic (modus ponens, modus tollens, contrapositio and so on) are valid. This is guaranteed by the adequacy of TFA:

(5.8) α is in TFT iff α is derivable by MP in finite number of steps from TFA.[XX]

Hence, every truth-functional tautology is a (SCI) tautology.

We call (5.5) the <u>special identity axiom</u>. The group (5.2), (5.3) (5.4) constitutes the <u>invariance axioms</u>. Observe that they are concerned with each (!) formator we have in the language.[XXI] Consequently, $\vdash$ deals with identity connective as genuine identity. Two replacement laws, salva identitate and salva veritate, are tautological:

(5.8) $\vdash [(\alpha \equiv \beta) \Rightarrow (\gamma(^{v}/_{\alpha}) \equiv \gamma(^{v}/_{\beta}))]$

(5.9) $\vdash [(\alpha \equiv \beta) \Rightarrow (\gamma(^v/\alpha) \Leftrightarrow \gamma(^v/\beta))]$.

We also have the symmetry and transitivity laws: $\vdash (\alpha \equiv \beta) \Leftrightarrow (\beta \equiv \alpha)$, $\vdash [(\alpha \equiv \beta) \wedge (\beta \equiv \gamma). \Rightarrow (\alpha \equiv \gamma)]$, and $\vdash (\alpha \not\equiv \neg\alpha)$.

The entailment $\vdash$, as every entailment, induces a collection of theories. A set X of formulas is here a theory iff X includes LA and contains β whenever it contains α and $\alpha \Rightarrow \beta$. We introduce some other notions which also apply to any entailment. First, we introduce a notation which comes from the theory of consequence operations. Given a set of formulas X, let TH(X) = the set of all formulas entailed by X. In other words, TH(X) = the smallest theory over X, i.e., the theory based on the axiom set X. Furthermore, we say that a set X is consistent if TH(X) $\neq$ FM = the set of all formulas. Maximal consistent sets are called complete and since they all are theories as a rule, we may talk of complete theories.

Properties of complete theories store much information on the underlying entailment. Here, complete theories are precisely those consistent theories T which satisfy: either α or $\neg\alpha$ is in T or (equivalently,) T includes either α or β whenever T includes $\alpha \vee \beta$. Finally, the entailment $\vdash$ is regular in the sense that the collection of all theories has the following separation property:

(5.10) if α is not in TH(X) then there exists a complete theory which does not include α but contains TH(X);(XXIII) obviously, the converse also holds.

6. Completeness and adequacy.

Three very distinct constructions sketched in sections 3,4 and 5 resulted in definitions of the entailments $\vdash^{(\underline{t})}$, $\vdash^{(\underline{r})}$, and $\vdash$. But, the truth is that we were dealing with the same thing. Indeed, one may prove the equalities:

(6.1) $$\vdash^{(\underline{\underline{t}})} = \vdash = \vdash^{(\underline{\underline{r}})}$$

Hence, the corresponding tautologies constitute exactly one set:

(6.2) $\vdash^{(\underline{\underline{t}})}$ tautologies = $\vdash$ tautologies = $\vdash^{(\underline{\underline{r}})}$ tautologies

Two equations in (6.2) are said, in the current terminology, to present two completeness theorems for SCI, considered as a formal system, i.e., set of formulas ($\vdash$ tautologies) generated from LA by MP.[(XXV)] On the other hand, (6.1) is said to present two so called generalized or strong completeness theorems for SCI[(XXVI)]

It is not my intention to combat definitions or change terminology for my own amusement. Yet terminological differences sometimes emphasize, clear cut differences in methodological approach. This is the case here and we decide to focus on equalities $\vdash = \vdash^{(\underline{\underline{t}})}$ and $\vdash = \vdash^{(\underline{\underline{r}})}$ in (6.1) and to call them <u>completeness theorems for the SCI</u>. The equations involved in (6.2) may be called weak completeness theorems for the SCI. We thus stress the fact that we are not thinking within the framework of the theory of formal systems and we do <u>not</u> concentrate on the set of theorems, i.e., formulas derivable from the axioms only in formal systems. Our theoretical framework is the general theory of entailment relations and, therefore, when facing a logical calculus we ask first for the entailment embodied in it and, what amounts to the same thing, for the associated totality of all theories based on that entailment.[(XXVII)]

Thus, in (6.1) we have two distinct completeness theorems for $\vdash$. Two distinct frameworks of semantical interpretations for SCI, the theory of truth-valuations and the theory of models, provide entailments $\vdash^{(\underline{\underline{t}})}$ and $\vdash^{(\underline{\underline{r}})}$ which equal $\vdash$. Let us comment for a while on that fact. You will fool yourself and the public if you claim any philosophical depth in preferring $\vdash^{(\underline{\underline{t}})}$ over $\vdash^{(\underline{\underline{r}})}$. To be sure, the machinery of truth-valuations and that of models are different and may present different difficulties in operating with them. But they are equivalent as far as two-valued and extensional logic is concerned. In fact, (6.1) holds even for the full NFL in comprehensive languages.[XXVIII]

Instead of doing falacious philosophy, ponder a little the equations (6.1). We have three distinct but equivalent definitions of an entailment. This circumstance strongly suggests (as in the theory of computable functions) that we are actually concerned with THE two-valued and extensional entailment. Subsequently, reflect that we have <u>two</u> distinct completeness theorems for the SCI. Thus, the well known relativity of completeness theorems is revealed, again. (I mean relativity to some semantical interpretation for the syntactically given logic.)

However, human yearning for the absolute cannot be extinguished. But equally eternal is human self-deception. Suppose, you have syntactically constructed in some sense, a deviant sentential logic.[XXIX] You have axioms AX and rules of inference RL. Hence, you have theorems, too. Then, you choose a semantical (say model-theoretic) interpretation SI (e.g., possible world semantics) and you prove the weak completeness theorem relative to SI. You are satisfied and think that SI provides an understanding of the logic you have built. But, you actually have very little, I assure you. To be sure, there exists another framework SI* and a corresponding weak completeness theorem for your logic. You even know it well but you disquality SI* as "unintended" and the cheating ceremony begins. First, one cannot speak about "intended" interpretations of a logic or particular theory unless its use is well established. For example, it is perfectly clear what the intended model of arithmetic is because we use it in elementary school. You know that the intended models of set-theory may evoke some doubts. But, if you tell me that SI constitutes the intended interpretations for the logic you have constructed and analysed, I will not believe you. I bet you are completely unable to use the constructed logic and you simply do not know what you intended with it. But, I may try to help you a little to better understand your own construction. Look closer at what you have actually done. Let $\vdash^{(y)}$ be the entailment generated by AX and RL; the letter "y" means "you". Furthermore, try to prove the completeness theorem for $\vdash^{(y)}$. You may discover that your favorite SI works or does not work at all. Anyway, look at the collection of all theories of $\vdash^{(y)}$, i.e., based on your logic. Then, an ingenious elaboration of Lindenbaum's ideas by R. Wojcicki [29] tells you that the language together with all your theories constitute a model-theoretic framework SI* with a completeness theorem for $\vdash^{(y)}$; (also see [30]). However, I am almost sure you are not satisfied with SI*. Nevertheless, you must be nonplussed, since any entailment and the collection of its theories are like head and tail of the same coin. So, SI* presents exactly what you actually intended by $\vdash^{(y)}$ or something very close to it. It is a fact that the collection of all theories of a given entailment looks very chaotic, in general. But, if your logic has some extensional properties (e.g., the existence of so called logical congruences) then you can factor SI* to SI** which is much simpler than SI* and which also provides you with a completeness theorem for $\vdash^{(y)}$ (see[25]). Moreover, if in addition, $\vdash^{(y)}$ is regular, then instead of SI** you may take only a special part of it SI_o and still have a completeness theorem for $\vdash^{(y)}$ (see [25]). This is the way one gets Lindenbaum-Tarski quotient models. Moreover, it is the only way to prove the completeness theorem (Post, Goedel) for Fregean logic. But, you are still unsatisfied with SI_o. After a while, you suddenly declare that I misunderstood you; the relation $\vdash^{(y)}$ is not the entailment you intended when specifying AX and RL. You patiently explain to me that certain rules in FL (e.g., the Goedel rule

$\alpha/\Box\alpha$ of necessitation) are admissible within the set of theorems and are not valid in any other sense. The axioms AX and rules RL define a formal system whose theorems are $\vdash^{(y)}$ tautologies but $\vdash^{(y)}$ is not the entailment you intended. Eventually, you give a syntactic definition of what you mean by entailment. However, you still wonder what use could be made of the new entailment since its tautologies are the same as those of $\vdash^{(y)}$.

So it is. To be sure, the Lord God will forgive you your attempt to fool me with the entailment $\vdash^{(y)}$. However, no one can save you from cheating yourself if you want to. So, listen. Use the Wojcicki method and carefully analyse the collection of all theories of your logic, i.e., the entailment you truly intend. This method always ends with a completeness theorem and is an adequate criterion of how good your logic is. In short, your logic is good only if Wojcicki's method leads to nice models. If your logic is bad then Wojcicki's method gives you a mess, naturally.

The relativity of completeness theorems is not only an opportunity for fruitless discussions of intended interpretations. In fact, the said relativity allows us to sharpen the completeness problem for a logic to what may be called the <u>adequacy problem</u> [8]. In fact, the SCI provides a good illustration. All models are involved in the completeness theorem: $\vdash \; = \; \vdash^{(\underline{r})}$. This theorem remains true if we restrict the class of models involved to countable ones. On the other hand, the class of finite models is too small, as previously noted. But, what about a class which contains just one, countable or uncountable model?

If $\vdash \; = \; \vdash_{\underline{M}}$ then the model $\underline{M}$ is called <u>adequate for</u> $\vdash$. Obviously, a model $\underline{M}$ adequate for $\vdash$ is also adequate for TAUT which means that TAUT = TR($\underline{M}$). The existence of a model, adequate for $\vdash$ - (TAUT) is a very strong form of the (weak) completeness theorem for SCI. Now, we know [14] that

(6.3) $\vdash$ has adequate models of the continuum power,

(6.4) TAUT has countable adequate models and, every such model is infinite.

Thus, one may naturally ask for countable adequate models for $\vdash$. The answer is [31]:

(6.5) any model, adequate for $\vdash$, is uncountable.

This must be seen as an extreme weakness of the SCI.

7. Non-Fregean logics.

The entailment $\vdash$ is very weak. Its weakness is expressed by several facts: the diversity of models, the uncountability of adequate models, validity of the natural postulate (3.11) and the great variety of theories. Indeed, genuine logic should be as weak as possible. I feel, however, that we might have some reasons to seek for a stronger entailment, i.e., an extension of $\vdash$. (XXX)

Again, the class of all (even finite and structural) extensions of $\vdash$ in the same language is enormously large. On the other hand, each such extension generates a collection of its own theories which is, in general, very large, also. Thus, we face an embarrassement of riches. Of course, one can think of extensions of $\vdash$ as syntactically defined and, then divide them into two classes: elementary and non-elementary extensions of $\vdash$. The former are defined as $\vdash$ with an additional (invariant) set of axioms added to LA. The later can be defined as $\vdash$ with some additional rules (schemes) of inference besides MP.

Consider, for example, the "G-rule", given by the scheme of inference

(7.1) $\alpha, \beta / \alpha \equiv \beta$

It is not $\vdash$ valid in view of (3.11) and determines a proper non-elementary extension $\vdash^G$ of $\vdash$. Sometimes another rule will be also called "G-rule". Let $\underline{1}$ stand for a fixed formula in TFT, e.g., $p \vee \neg p$ and write $\Box\alpha$ for the equation $\alpha \equiv 1$. Then it is easily seen that the rule $\alpha / \Box\alpha$ is equivalent to the G-rule (7.1). This fact explains our terminology and, perhaps, creates an illusion that we aim at modal, intensional logic. Please, do not think so.(XXXI) Rather observe, in this connection, that

(7.2) there exists no SCI theory T such that: $\Box\alpha$ is in T if and only if α is in TAUT.

However, there are plenty of theories T with the property: $\Box\alpha$ is in T iff α is in T. They are precisely the theories of $\vdash^G$ and we call them "G-theories".(XXXII)

As far as elementary extensions of $\vdash$ are concerned, we may only consider those entailments $\vdash_T$ which are generated inductively by MP and LA $\cup$ T where T is any theory(!). Since T = $TAUT_T$ = tautologies of $\vdash_T$ we will, naturally choose T to be an invariant theory. So we discuss invariant theories for a while.

If X is a set of formulas then Sb(x) = the set of all those formulas which result from formulas in X by the substitution of formulas for variables. We write Sb(α) instead of Sb($\{\alpha\}$). Invariant theories are precisely those theories T such that T = TH(Sb(X)) = the invariant theory based on the axiom set X, for some subset X of FM.(XXXIII)

For example, consider the invariant theory WF based on a single formula:

(7.3) $(p \equiv q) \vee (p \equiv r) \vee (q \equiv r)$.

It is essentially well known that WF, the Fregean logic, is the only consistent invariant theory of $\vdash_{WF}$, i.e., WF is "Post-complete". Hence, $\vdash_{WF}$ is an elementary extension of $\vdash$ with a certain definite property of maximality.

Non-elementary extensions of $\vdash$ are not "classical" in the sense of Bloom and Brown, [32]. Consequently, their semantics is rather strange and hard to develop. To be sure, it is not difficult to have corresponding completeness theorems. However, questions concerned with adequacy have mostly negative answers. For example, there is no model M such that $\vdash^G = \vdash_M$.

On the other hand, all elementary extensions of $\vdash$ are classical and have nice semantical properties. I only mention the adequacy problem. A model M is adequate for $\vdash_T$ (or for T) where T is an invariant theory if, correspondingly, $\vdash_T = \vdash_M$ (or $T = TR(M)$). Then, we have the following:

(7.4) $\vdash_T$ has an adequate model iff T has an adequate model.(XXXIV)

Thus, given an invariant theory T, we are interested in a necessary and sufficient condition for the existence of adequate models for $\vdash_T$ or, equivalently, for T. It is very pleasant to know that the desired condition is so called <u>quasi-completeness</u>, a very old notion of Łos; see [33] and [34]. It is better known in the theory of modal systems as <u>Hallden-completeness</u> (or reasonableness), [35]. A consistent invariant theory T is called <u>quasi-complete</u> iff the following holds: if no variable occurs both in α and β and $\alpha \vee \beta$ is in T then either α or β in T. Then, we have the Łos theorem:

(7.5) T has an adequate model iff T is quasi-complete.

This theorem is also true for NFL in open W-languages with sentential and nominal variables; see [13]. Again, we have the analogy as mentioned in section 1.(XXXV)

The entailment $\vdash_{WF}$ is called the Fregean logic in the given W-language (here, SCI-language) and will be discussed in the next section. All other (nontrivial) extensions of $\vdash$ are called non-Fregean logic. We may also talk of Fregean and non-Fregean theories. A theory is called Fregean if it contains WF or, equivalently, $Sb(\alpha_F)$. Otherwise, the theory is non-Fregean.

As noted previously, the totality of all non-Fregean logics is enormously large and diverse. Hence, to narrow the field of investigation by making certain choices seems to be an absolute necessity. Of course, we may disregard all non-elementary extensions of $\vdash$ and, subsequently, consider only those elementary extensions $\vdash_T$ of where T is an invariant theory. But, even this class is tremendously large. There exist, for example, infinitely many entailments $\vdash_T$ whose tautologies T constitute a Post-complete theory, distinct from WF.(XXXVI) Therefore we want to restrict the class of non-Fregean logics once again. We feel strongly that the Fregean logic is a particularly distinguished extension of $\vdash$ and, we should make use of it some way. Thus, as the first move, we decide to only consider entailments weaker than $\vdash_{WF}$. In other words, we focus our attention upon the interval between $\vdash$ and $\vdash_{WF}$. Thus, the logics we are going to consider from now on are entailments $\vdash_T$ where T is any invariant theory contained in WF. Then, $\vdash_T$ is non-Fregean whenever T does not contain WF, that is, $T \neq WF$ and, $\vdash_T = \vdash$ iff $T = TAUT$.

Since $\vdash$ is the pure (absolute) non-Fregean logic, it follows that each non-Fregean logic (of the restricted kind) distinct from $\vdash$ involves certain non-logical content, that is, some non-tautological assumptions which we call ontological. Clearly, the same applies to the Fregean logic $\vdash_{WF}$ which is the greatest one in our interval of entailments. Hence, by studying non-Fregean logics, weaker than $\vdash_{WF}$ we may hope to reveal and analyse the ontological content of the Fregean logic.

The Fregean logic is determined by the invariant theory WF based on single axiom α_F, i.e., the formula (7.3):

$$\text{(7.6)} \qquad WF = TH(Sb(\alpha_F))$$

In the next section we find other formulas which may serve, like α_F, as single axioms of WF. Later, we will select three particular formulas α_B, α_T and α_H and, discuss three invarient theories based on them as single axioms: WB, WT and WH. These theories are sets of tautologies of three non-Fregean logics: $\vdash_{WB}$, $\vdash_{WT}$ and $\vdash_{WH}$.

Each logic $\vdash_{WZ}$ where Z is B, T, H or F is classical and, hence, has both the finiteness and deduction properties. Therefore, $\vdash_{WZ}$ reduces, in a sense, to the set of its own tautologies, WZ. Now, WZ is an invariant theory. For each theorem α in WZ the whole set $Sb(\alpha)$ is contained in WZ. So, we are allowed to say that ontological assumptions involved in WZ and underlying $\vdash_{WZ}$ are universal statements concerned with the totality of situations and operations on them.

8. Fregean logic.

Since WF is an invarient theory we may use the substitution rule within it. Hence, formal derivations easily lead to the following theorems of WF:

$$\text{(8.1)} \qquad (p \not\Leftrightarrow q) \Rightarrow (p \equiv q)$$

$$\text{(8.2)} \qquad (p \Leftrightarrow q) \Leftrightarrow (p \equiv q)$$

$$\text{(8.3)} \qquad (p \Leftrightarrow q) \equiv (p \equiv q)$$

We will refer to all four formulas (7.3), (8.1), (8.2), (8.3) ambiguously as α_F since they are all equivalent in the sense that they may serve as single axioms for WF. We call the formula α_F and the set $Sb(\alpha_F)$ the _ontological Fregean axiom_ and _(ontological) Fregean axiom scheme_, respectively, in contrast to the semantical or metatheoretical version of Fregean axiom stated at the beginning of this paper.

Clearly α_F is neither in TAUT nor in TFT and you may contemplate the superficial meanings of different versions of the Fregean axiom α_F. The essential point is, however, that the Fregean model M_F is the only one in which α_F is true. Moreover, since WF is post-complete

(8.4) $$WF = TR(M_F)$$

The model M_F is adequate not only for the theory WF but also for the Fregean entailment, i.e.,

(8.5) $$\vdash_{WF} = \vdash_{M_F}$$

It seems to be instructive to compare (8.5) and (6.5).

The variant (8.3) of the Fregean axiom identifies the identity connective with the material equivalence connective. Hence they are completely indistinguishable in the Fregean logic. In every Fregean theory, $\equiv$ is truth-functional and $\Leftrightarrow$ has all properties of identity connective. Therefore, we do not lose anything if we throw one of them out of the language. (Which one?) Anyway, if we so restrict the language we obtain the usual Fregean logic (FL) in sentential language, that is, the well-known truth-functional logic.

9. Three non-Fregean theories.

The three previously announced theories WB, WT and WH are invariant theories based on single axioms. Thus, $WZ = TH(Sb(\alpha_Z))$ where Z stands B,T or H and, the axioms α_B, α_T and α_H are defined in turn below.

The formula α_B is the conjunction of the following equations:

(9.1) $[(p \wedge q) \vee r] \equiv [(q \vee r) \wedge (p \vee r)], \quad [p \vee (q \wedge \neg q)] \equiv p,$

(9.2) $[(p \vee q) \wedge r] \equiv [(q \wedge r) \vee (p \wedge r)], \quad [p \wedge (q \vee \neg q)] \equiv p,$

(9.3) $(p \Rightarrow q) \equiv (\neg p \vee q), \quad (p \Leftrightarrow q) \equiv [(p \Rightarrow q) \wedge (q \Rightarrow p)].$

This is a short axiom system for Boolean algebras (see [36]), written with sentential variables and truth-functional connectives $\neg$, $\wedge$, $\vee$, $\Rightarrow$, $\Leftrightarrow$ as representing operations of complement, meet, join, co-difference and symmetric co-difference. Accordingly, all theories containing $Sb(\alpha_B)$ are called <u>Boolean</u>.

Let us briefly reflect on theorems of the smallest Boolean theory WB. First, we can prove the familiar equational Boolean laws: commutative, distributive and absorption laws, De Morgan laws and the double negation law. Furthermore, since we have the theorems $(p \vee \neg p) \equiv (q \vee \neg q)$ and $(p \wedge \neg p) \equiv (q \wedge \neg q)$ we may introduce two sentential constants $\underline{\underline{1}}$ and $\underline{\underline{0}}$ in two legitimate equational definitions:

(9.4) $\underline{\underline{1}} \equiv (p \vee \neg p) \qquad \underline{\underline{0}} \equiv (p \wedge \neg p)$

Then, $\underline{\underline{1}}$ and $\underline{\underline{0}}$ are the unit and zero of the Boolean algebra of situations. This fact can also be expressed in terms of the ordering $\lesssim$ which we define, together with $\Box$, as follows: $\Box p \equiv (p \equiv 1)$; $p \lesssim q \equiv \Box(p \Rightarrow q)$. Thus:

(9.6) $\vdash_{WB} (p \lesssim q) \Rightarrow (\Box p \Rightarrow \Box q), \quad \vdash_{WB} \Box(p \wedge q) \Leftrightarrow [\Box p \wedge \Box q]$

(9.7) $\vdash_{WB}$ $(p \equiv q) \Leftrightarrow \Box(p \leftrightarrow q)$, $\vdash_{WB}$ $(p \leq q) \Leftrightarrow [(p \wedge q) \equiv p]$.

If you are thinking that WB is the theory of the Boolean algebra of situations then I am sure you have mixed feelings. First you are probably trying to understand the theorems of WB according to the conceptual framework of intensional necessity and strict implications. I advise you simply to forget it. You may, of course, read formulas of the form $\Box\alpha$ as "it is necessary that α" but you must necessarily purify the meaning of the word "necessity" from any intensional shadows. In Boolean theories, the formula $\Box\alpha$ means, by definition, exactly that (the situation that) α is the unit of the Boolean algebra of situations.

Obviously, the connective $\Box$ is here extensional, as is every formator in SCI (and NFL, in general) and you may rightly conclude that the ambiguous notion of necessity, used in natural language, has an extensional dimension, subject to the laws of Boolean theories in SCI.

But, even if you get rid eventually of all intensional ghosts the theory WB as a theory of certain Boolean algebra seems strange to you. No wonder! You have just come to another logical country and you are experiencing something quite new. Remember what I promised you at the beginning: analogy and difference. If you examine some theorems of WB you must agree that we deal in WB with some Boolean algebra. This is also said in the following semantical theorem: for any <u>model</u>(!) $M = \langle A, F \rangle$

(9.8) M is a model of WB, i.e., WB is contained in TR(M) iff A is a Boolean algebra (with circle operation).

Thus, WB actually is a theory of certain Boolean algebras and this is a case of analogy, again. However, some differences must appear since WB is written with sentential variables. Indeed, the formula

$\Box p \Rightarrow p$ is in TAUT. Hence, it is in WB. Therefore, and inference

$$\frac{\Box \alpha}{\alpha}\ , \text{ i.e., } \frac{\alpha \equiv \underline{\underline{1}}}{\alpha} \text{ hence } \alpha$$

is here perfectly legitimate. But nothing like that makes any sense in the theory of ordinary Boolean algebras.(XXXVII)

Now, we pass to the theory WT. The formula α_T is the conjunction of α_B and four equations:

(9.9) $\Box(p \Leftrightarrow q) \equiv (p \equiv q)$, $\Box p \leqslant p$,

(9.10) $\Box(p \wedge q) \equiv [\Box p \wedge \Box q]$, $\Box\Box p \equiv \Box p$.

Obviously, every theorem of WT is a theorem of WB but not conversely. We see that the connective $\Box$ represents in WT a kind of topological interior operator on the Boolean algebra of situations. Hence, granted WT, we may talk of the topological Boolean algebra of situations.(XXXVIII) Here is a sample of theorems of WT which are not in WB:

(9.11) $(\Box \underline{\underline{1}}) \equiv \underline{\underline{1}}$, $(p \equiv q) \equiv \Box(p \equiv q)$, $(p \equiv q) \leqslant (p \Leftrightarrow q)$,

(9.12) $(p \equiv q) \equiv (q \equiv p)$, $[(p \equiv q) \wedge (q \equiv r)] \leqslant (p \equiv r)$,

(9.13) $(p \leqslant q) \equiv [(p \wedge q) \equiv p]$, $[\Box p \equiv \neg\Box\neg p] \Rightarrow [(p \equiv \underline{\underline{1}}) \vee (p \equiv \underline{\underline{0}})]$

Notice that $\Box$ and $\equiv$ are interdefinable in WT. Therefore, additional axioms (9.9) and (9.10) are, in fact, concerned with the identity connective. For other details see [9] and [37].(XXXIX)

Both theories WB and WT are non-Fregean. If we add to them the formula (another version of the Fregean axiom)

(9.14) $(p \equiv \underline{1}) \vee (p \equiv \underline{0})$

we obtain WF. But, what about apparently weaker formula

(9.15) $[(p \equiv q) \equiv \underline{\underline{1}}] \vee [(p \equiv q) \equiv \underline{\underline{0}}]$

This formula is characteristic for the Boolean theory WH. In fact, its single axiom α_H, is the conjunction of α_B and (9.15). Another equivalent version of α_H is the conjunction of α_T and the formula

(9.16) $[(\Box p) \equiv \underline{\underline{1}}] \vee [(\Box p) \equiv \underline{\underline{0}}]$.(XL)

Thus, WH is much stronger than WT and its content is that the topological Boolean algebra of situations is a <u>Henle algebra</u>. Indeed, for any <u>model</u>(!) $\underline{\underline{M}} = \langle \underline{\underline{A}}, F \rangle$

(9.17) M is a model of WH iff A is a Henle algebra.

It follows that WH is also a non-Fregean theory.

WB, WT, WH and WF are invarient theories listed in order of increasing strength. All of them are equational theories, that is, of the form TH(X) where X is a set of equations. Properties of WF are well-known. On the other hand, I do not have much to say here about WB. So, I only discuss certain properties of theories WT and WH.

First of all, both theories WT and WH are G-theories and, in this respect, the theory WT is the smallest G-theory over WB. Consequently,

(9.18) the following three conditions are equivalent for any T:

(a) T is an equational theory over WT,

(b) T is a G-theory over WB,

(c) T is a theory such that for all formulas α, β :

$(\alpha \equiv \beta)$ is in T iff $(\alpha \Leftrightarrow \beta)$ is in T.

In general, every G-theory is equational but not conversely. For example, no equational theory over WB, properly contained in WT is a G-theory.

Furthermore, one may easily construct uncountable models, adequate for WT or WH. Hence, both theories are quasi-complete and both entailments $\vdash_{WT}$ and $\vdash_{WH}$ have adequate models. However, there seems to be a big difference between WT and WH.

(9.19) $\vdash_{WH}$ has a countable infinite adequate models.

On the other hand, $\vdash_{WT}$ seems to be like $\vdash$ as far as adequate models are concerned. I conjecture that every model adequate for $\vdash_{WT}$ is uncountable. This is suggested by several minor differences between WT and WH which may, however, be omitted here.

Entailments $\vdash_{WT}$ and $\vdash_{WH}$ belong to the interval between $\vdash$ and $\vdash_{WF}$. If we jump from $\vdash_{WF}$ to $\vdash$ then we lose many valuable properties the Fregean logic has. There is nothing wrong with that. Quite to the contrary. We simply learn that $\vdash_{WF}$ is not the pure logic and that we owe much to the ontological Fregean axiom. In general, every proper (structural) elementary extension of $\vdash$, including $\vdash_{WF}$, is the result of contaminating the pure logic with some ontological principles concerning the structure of the universe of situations. Clearly, one may raise a philosophical problem immediately. Suppose, we choose an ontological principle α_Z (or, $Sb(\alpha_Z)$ rather) where Z is B,T,H or F. Are we discovering thereby the real structure of the universe of situations or only making a convention to please ourselves some way? But do not hurry to answer this question. Any answer, including that one firmly preferred by myself, does not matter here. Another question seems to be primary. The Fregean logic is based on extremely strong ontological principle (the Fregean axiom) and has well known nice properties. So, we shall ask first which properties of a logic $\vdash_T$ are induced by ontological principles accepted in T. For example, consider the replacement property of logically equivalent formulas: "logically equivalent formulas are interchangable in all contexts salva veritate" as Quine put it in [12], p. 159. Well, interchangability is an intrinsic property of identicals and holds for $\vdash$ and any extension of it. Why must it be shared by logical equivalents? Because of the Fregean axiom, the replacement property of logically equivalent formulas holds in Fregean logic. It clearly fails for $\vdash$ by the natural postulate (3.12). But consider an entailment $\vdash_T$ where T is an invariant theory. Two formulas α and β are logically equivalent with respect to $\vdash_T$ iff $\alpha \Leftrightarrow \beta$ is a tautology of $\vdash_T$, i.e., in T. Thus, we see by (9.18) that if T is any (invariant) equational theory over WT then $\vdash_T$-logically equivalent formulas are interchanable in all contexts salva identitate and salva veritate. That is, if $\alpha \Leftrightarrow \beta$ is in T then all formulas of the form

$$\gamma[^v/\alpha] \equiv \gamma[^v/\beta] \quad \text{or} \quad \gamma[^v/\alpha] \Leftrightarrow \gamma[^v/\beta]$$

are theorems of every theory over T. This is an example of how some nice properties of $\vdash_T$ depend on ontological assumptions in T.(XLI)

10. SCI and modal systems.

It came as a surprise to me when I discovered in 1969 that

(10.1) the SCI-theory WT equals Lewis system S_4 and, consequently, every modal system containing S_4 is an SCI-theory; in particular, S_5 = WH.(XLII)

Now, you are pretty sure that SCI is a kind of modal logic. But, my response is still an emphatic NO. Unfortunately, the divergence in our opinions cannot be resolved by formal reasoning. You know exactly what non-Fregean logic is. But, you do not have any precise and adequate idea about what modal logic is. Do you? The best I can do in defense of my point is to continue to argue for extensionality and logical two-valuedeness of SCI and, in particular, to show that it does not fit the intensional whims of modal logic.

First of all, one may try to consider modal logic as an entailment imposed on a language. However, the trouble is that modalists and other intensionalists talk much about entailment but almost never make any effort to define it. So, I suppose that they are not actually interested in entailment and, hence, do not know what they are talking about. Formerly, I thought in [9] that modal entailment could be defined as certain extension of $\vdash^G$. Now, I see how wrong I was. Too many intensional ghosts surround modal entailment so that one cannot be sure whether it exists at all. With regret, I no longer consider modal logic as an entailment.(XLIII)

However, modal systems exist as well defined <u>sets of formulas</u> and we may attempt to compare them with theories of SCI. The idea is to equate the strict equivalencies defined as $\Box(\alpha \Leftrightarrow \beta)$ with equations $\alpha \equiv \beta$. This is a kind of syntactic translation of formulas into formulas and it also leads to the result (10.1) above. As far as weaker modal systems are concerned we have the following result.

(10.2) All modal systems contained in the Feys-von Wright system S_o are not SCI-theories. Also, every modal system containing S_o, properly contained in S_4 and closed under necissitation rule (G-rule) is not SCI-theory. In particular, the Brouwersche system is not SCI-theory.

In view of (10.2), you may ponder (10.1), again. You certainly know that S_4 and S_5 have been introduced in 1930 by O. Becker and are not Lewis systems, strictly speaking. Lewis preferred weaker modal systems. Thus, stronger modal systems are examples of how an intensional mess may, unconsciously and rather involuntarily, give birth to theories of extensional logic. And I cannot help wondering whether Scott in [1] defends Lewis' program or his own.

There is a single reason for ruling out in (10.2) so many modal systems as non-theories: none of these systems contain (5.4) formulated with strict equivalence in place of identity. Thus, although the strict equivalence connective behaves, in S_4 and stronger modal systems, as an extensional identity connective, in almost all other modal systems it is not extensional with respect to itself. It was probably intentionally intended to be intensional.

Lewis system S_3 may present, possibly, an exception. However, if S_3 actually is an invariant Boolean theory then you will find that the formula $\Box(p \equiv p)$ is not in it. This may puzzle you a little. But why to bother about S_3 at all? Rather forget it since you may directly study all and every invarient theory in the interval between WB and WT and, corresponding models. Notice that some of those theories contain the formula $\neg \Box(p \equiv p)$. These theories can be easily extended to Post-complete theories which are, however, necessarily non-Fregean. Recall that $\Box(p \equiv p)$ is in WF.(XLIV)

Alas! one cannot kill the modal logic by shooting it with formal arguments as above. Probably, modal logic cannot be killed at all. Fortunately, we do not need any killing. What we really need is a complete separation of the non-Fregean logic from the modal logic.

So, we are going to show that certain intensional desires, typical for modal logic, are flatly denied by NFL.[XLV]

11. Notes on intensional necessity.

Belief-sentences are paramount to intensional contexts. Someone may believe that α and not believe that β although $\alpha \equiv \beta$. Clearly, belief-contexts do not fit to NFL. You may easily reconstruct the Morning-Evening-Star paradox with belief-sentences and see that they violate replacement laws of equals by equals. The conclusion is that there is no functional dependence of

(11.1) (the situation) that x believes that p

on

(11.2) the person x and (the situation) that p.

How (11.1) depends on (11.2) I do not know.

Similar difficulty is provided by the intensional notion of necessity. Furthermore, you will clearly see that this notion is not represented in SCI by the square connective $\Box$. But, first let us discuss some properties of equations. Now, we consider open W-languages with two kinds of equations $\varphi \equiv \psi$ where $\equiv$ is either the identity connective or identity predicate.[XLVI]

Suppose that T is a consistent invarient Boolean theory containing all formulas of the form:

$$\Box(\varphi \equiv \varphi) \tag{11.3}$$

Then, the theory T contains all formulas of either form:

(11.4) $(\varphi \equiv \psi) \Leftrightarrow \Box(\varphi \equiv \psi)$

(11.5) $(\varphi_1 \equiv \psi_1) \wedge (\varphi_2 \equiv \psi_2) \Rightarrow (\varphi_1 \equiv \psi_1) \equiv (\varphi_2 \equiv \psi_2)$

(11.6) $(\varphi \equiv \varphi) \equiv \Box(\varphi \equiv \varphi)$

Thus, granted T, we may roughly say that as far as equations are concerned all and only all true equations describe the same situation, vis., the unit of the Boolean algebra of situations. You may see it also in models and everything is perfectly all right if only you do not listen to the intensional ghosts. They whisper to you that there is something wrong with (11.4), that (11.4) could be acceptable only if restricted to analytic equations and you had better get rid of NFL and resort to some impossible world. But, I tell you: do not make the slightest concession to intensional shosts if you want to keep an extensional position. Remember the distinction between reference and sense and do not confuse what you are speaking about with how you are speaking about it. This kind of confusion is very close to the use-and mention confusion. An analytic sentence and a non-analytic true sentence may refer to the same. There is nothing wrong with this. You know what formulas of the form $\Box\alpha$ mean in SCI. They do not mean anything like "the formula α is analytic." Moreover, I will show that the square connective $\Box$ cannot imitate the metatheoretic predicate of analyticity according to the following postulate:

(11.7) $\Box\alpha$ is true iff α is analytic.

Whatever analytic formulas are, they certainly constitute a consistent invariant theory T. Then, the synonymity relation relative to T, $\varphi \underset{T}{\sim} \psi$, is defined as

(11.8) the equation $\varphi \equiv \psi$ is in T.

Since $\underset{T}{\sim}$ depends on equations in T only it follows that two distinct theories may define the same synonymity relation. What determines $\underset{T}{\sim}$ is the largest equational theory contained in T, called the <u>equational kernel</u> of T. Obviously, T = equational kernel of T if and only if T is equational itself. Hence, we suppose additionally that the theory T, i.e., the set of analytical formulas, is an equational theory. The remaining assumptions on T are concerned with how analycity is related to truth. A truth-valuation t is said T-admissible if

(11.9) $t(\alpha) = 1$ whenever α is in T and,

(11.10) $t(\varphi \equiv \psi) = 1$ for some equation $\varphi \equiv \psi$ <u>not</u> in T.

Thus, we suppose that T-admissible truth-valuations exist and only these will matter in our discussion. Hence, the problem whether the postulate (11.7) can be realised reduces to the following question. Given a set T of analytical formulas, as above, does there exist a T-admissible truth-valuation t such that

(11.11) $t(\Box\alpha) = 1$ iff α is in T.

We consider three cases: TAUT, WT and WH. Suppose, T = TAUT. Then, by (7.2), there is no truth-valuation t such that (11.11). Thus, in this case the postulate (11.7) cannot be realized. Passing to WT and WH, recall that both are G-theories. If T is WT or WH then clearly there exist truth-valuations t such that (11.9) holds. However, each of them satisfies either (11.10) or (11.11) but not both. Moreover, if T = WH then for every truth-valuation t such that (11.9) the condition (11.10) is satisfied but (11.11) is not. Again, the postulate (11.7) is denied.

Quite similar reasoning will show that the ordering connective

$\leqslant$ cannot imitate entailment. Ask whether there exists a T-admissible truth-valuation t such that

(11.12) $$t(\alpha \leqslant \beta) = 1 \text{ iff } \alpha \vdash_T \beta$$

and suppose that T is WT or WH. You may also take TAUT for T. But, then the connective $\leqslant$ is hardly to be called an ordering connective.

Finally, you may verify yourself that the identity connective is no imitation of either logical equivalence or synonimity relation.

Let me conclude this chapter with some remarks related to the synonymity relations $\sim_T$. Obviously, they lead immediately to equivalence classes representing senses of expressions modulo T and prove to be very useful in model theory of NFL. But this point may be omitted here. Rather, we will jump to strong equational theories. Actually, we do not depart very far from synonymity relations since they are in a one-one correspondence with equational theories. We want to point out a particular property of some theories over certain G-theories which may play a role in the discussion of how NFL and FL are related.

Consider a theory T over WT.[(XLVII)] If T contains all formulas of the form

(11.13) $$(\alpha \equiv \underline{1}) \vee (\alpha \equiv \underline{0})$$

then T = WF and, conversely. We say that all formulas are zero-one valued relative to WF and define α to be a zero-one-valued formula (shortly, ZOV-formula) relative to T iff the disjunction (11.13) is in T.

First, it is easy to see that:

(11.14) T is equational (or, G-theory) if and only if every theorem of T, i.e., formula in T is a ZOV-formula relative to T.

Subsequently, consider the set of all ZOV-formulas relative to T which are theorems of T and call it the Fregean kernel of T. It is a G-theory and is the largest G-theory contained in T. Therefore,

(11.15) Fregean kernel of T = equational kernel of T.

Thus, the Fregean kernel of T is what determines the synonymity relation $\sim_T$.

12. Possible world semantics and SCI.

One may also shift to semantics and try to point out differences between extensionality of SCI and intensionality of modal logic. Consider for example, possible-world semantics for sentential modal systems only.(XLVIII)

The simple possible-world semantics, sketched on the first two pages of [42], works in the sense of a weak completeness theorem for some stronger modal systems only. If applied to SCI it only works in this sense for invariant Boolean G-theories. The reason here is similar to that for (10.2). The universe of simple Kripke models the sentential variables run over is the power set 2^I considered as Boolean algebra and its the unit serves as the only distinguished element. Moreover, as shown by the late D. Penkov in 1971, the logical axiom (5.4) of SCI forces the relation on the set I in Kripke models to be transitive.

Certainly, the intensional background of possible-world semantics is very complicated and difficult to describe and analyse. However, at least one idea of possible-world semantics for modal systems can be distinctly identified. It is the requirement to have just one distinguished element in the model, viz., the unit of the Boolean algebra. Of course, I am unable to tell what intensional ghost is responsible for this. Yet, I call it the semi-Fregean postulate, since it may be seen as the result of semi-abolition of the Fregean axiom, accomplished to pleast only some intensional ghosts.

Validity of the semi-Fregean postulate is clearly seen in the complicated possible-world semantics, designed for weak modal systems, and also in Lemmon's modal algebraic semantics [43]. Moreover, it is responsible for the significant fact that methods of modal semantics fail for those (invariant) Boolean theories which are not G-theories. To see this consider non-trivial modal algebras in the sense of [44]. These are non-trivial Boolean algebras $(0 \neq 1)$ supplemented with an additional unary operation I. To adapt modal algebras to SCI define the binary circle operation as follows:

(12.1) $$a \circ b \;=_{DF}\; I(a \div b)$$

Then, $Ia = a \circ 1$ and we see that a (Boolean) ultrafilter F of a modal algebra A is an f-set of A iff

(12.2) Ia is not in F whenever $a \neq 1$.

Obviously, not every modal algebra contains f-sets. But, if it does then it fits a partial semantics of SCI. In fact, the class of all models $\langle A,F \rangle$ where A is a modal algebra provides the completeness theorem for the entailment $\vdash_{WB^*}$ where WB^* is the smallest Boolean theory containing all formulas of the form:

(12.3) $$(\alpha \equiv \beta) \equiv \Box(\alpha \Leftrightarrow \beta)$$

Again, a brief comparison of (12.1) and (12.3) is a strong argument for the extensionality of SCI and its theories. It is also a good illustration of the difference between semi-abolition of the Fregean axiom which is an intensional affair and the extensional but full aboliton of this axiom.(XLIX)

It is true that ultrafilters of modal algebras are taken into account in modal semantics, sometimes; for example, in connection with Hallden completeness of modal systems (see [39], [45]). But the concept of an f-set is completely alien to intensional semanticists. It is a proper blend of two-valuedness and identity and, a genuine generalization of the unit of the two-element Boolean algebra (Fregean model). Although the use of f-sets annuls the semi-Fregean postulate of modal logic it may lead to interesting, perhaps, algebraic results even in the field occupied as yet by modal logicians. Certainly, f-sets of a modal algebra form, if they exist, a complicated class. But, suppose, we know well the structure of the class of all filters(!) F in modal algebras with the property (12.2). Then, we may raise a problem similar to that of Lemmon [43] and Makinson [44]. Is it possible to define a class of (relational) Kripke models which correspond, in an appropriate sense, to SCI-models based on modal algebras?

Thus, one may try to modify and, possibly enrich the possible-world semantics by explicit use of f-sets in Boolean algebras with a circle operation. Surprisingly enough, the converse procedure also seems to be possible. M.J. Cresswell actually modified in [46] the possible-world semantics so that it fits the whole SCI in the sense of completeness theorem. This is very embarrasing since it weakens the methodological thesis about NFL as an extensional logic, essentially distinct from modal logic. However, a detailed analysis of Cresswell models, carried out in collaboration with Stephen L. Bloom, hows that Cresswell's remarkable result is completely misleading and an example of how formal, set-theoretical constructions can support superficially a deceiving philosophy.

13. Cresswell models for SCI.

To simplify the discussion of Cresswell models, only the negation and conjunction will be explicitly mentioned as truth-functional connectives of the SCI-language.

Given a non-empty set I, the powerset 2^I is the set of all functions from I to $\{0,1\}$. If f is in 2^I and j in I then the value of f at j will be denoted as f_j or, in complex contexts, $p_j(f)$. For each j in I we form the set $H^{(j)}$ of all functions f in 2^I such that $f_j = 1$.

Let G_1, G_2, G_0 be, correspondingly, a unary and two binary operations on 2^I. Thus, for all f and g in 2^I, 2^I also contains $G_0(f,g)$, $G_1(f)$ and $G_2(f,g)$. Subsequently, the algebra

(13.1) $$G = \langle 2^I, G_0, G_1, G_2 \rangle$$

is called a c-algebra on I and the structure

(13.2) $$C = \langle G, H^{(j)} \rangle = \langle 2^I, G_0, G_1, G_2, H^{(j)} \rangle$$

is said to be a c-model on $\langle I,j \rangle$ iff the following conditions are satisfied for all f and g in 2^I:

(13.3) $$p_j(G_0(f,g)) = 1 \text{ iff } f = g$$
$$p_j(G_1(f)) = 1 \text{ iff } f_j = 0$$
$$p_j(G_2(f,g)) = 1 \text{ iff } f_j = g_j = 1$$

Cresswell models are represented here as c-models.[LI] Elements of $H^{(j)}$ play the role of distinguished elements in c-models. If h

is an homomorphism from the SCI-language to the c-algebra G then we say that h satisfies a formula α in the c-model C as in (13.1) if $h(\alpha)$ is in $H^{(j)}$. Let $St_h(C)$ be the set of all formulas satisfied by h in C. Then, we have the Cresswell theorem:

(13.4) $X \vdash \alpha$ iff for every non-empty set I, any j in I and every c-model C on $\langle I,j \rangle$, if for some homorphism h the set X is contained in $St_h(C)$ then also α is in $St_h(C)$.

This is a completeness theorem for SCI. To analyse it, we want to find some relationship of c-models to our SCI-models. First, we show how to obtain certain c-models from SCI-models. Let $M = \langle A,F \rangle$ be a model and Q be a non-empty family of subsets of the universe A of M. Define the map D from A to 2^Q as follows. For every X in Q and every a in A,

(13.5) $$p_X(D^{(a)}) = 1 \quad \text{iff} \quad a \text{ is in } X.$$

Here, $D^{(a)}$ is the value of D at a. Then call Q admissible if Q has two properties:

(13.6) $$F \text{ is in } Q \text{ and,}$$

(13.7) $$D^{(a)} \neq D^{(b)} \quad \text{whenever} \quad a \neq b.$$

The condition (13.7) is a kind of separation property imposed on Q. If Q is admissible then D is one-one and embeds A into 2^Q. Consequently, the operations $-$, $\cap$, $\circ$ of A define operations G_1, G_2, G_0 on the f-image of A in 2^Q, denoted as $D(A)$. Furthermore, operations G_1, G_2, G_0 may be extended arbitrarily(!) over the whole powerset 2^Q except that we require the following: for all f,g in 2^Q

(13.8) $$p_F(G_1(f)) = 1 \text{ iff } f_F = 0$$

$$p_F(G_2(f,g)) = 1 \text{ iff } f_F = g_F = 1$$

$$p_F(G_0(f,g)) = 1 \text{ iff } f = g$$

Finally, let $H^{(F)}$ be the set of all g in 2^Q such that $g_F = 1$.

It is immediately seen that the structure

(13.9) $$C(M) = \langle 2^Q, G_0, G_1, G_2\ H^{(F)} \rangle$$

is a c-model on Q,F. The notation "$C(M)$" neglects the admissible set Q. Obviously, the class of admissible sets presents a great variety. For example, Q may be of the same cardinality as A. In the extreme case, $Q = P(A)$, the set of all subsets of A. But, however we chose an admissible Q the c-model $C(M)$ contains a submodel whose universe is $D(A)$, called the kernel of $C(M)$. It is not a c-model but an SCI-model, isomorphic to the original model M. Hence, some semantical properties of M are inherited by $C(M)$. In fact, if r is a valuation over M (homomorphism of the SCI-language to the algebra A) then put $\bar{r}(\alpha) = D^{(r(\alpha))}$. The map $\bar{r}$ sends formulas into $D(A)$ and, hence, into 2^Q and, is a homorphism to the algebra of $C(M)$. We say that $\bar{r}$ is induced by r and we have the following basic equality:

(13.10) $$St_{\bar{r}}(C(M)) = St_r(M)$$

where $St_r(M)$ is the set of formulas satisfied by r in M.

A c-model, obtained as $C(M)$ from some SCI-model M will be called special.[LII] These c-models involve SCI-models as kernels and, hence, one may hope that they suffice to some extent in semantics of SCI. Indeed, a close inspection of the Cresswell proof of his theorem shows that the class of special c-models (of the cardinality of continuum at most) provides the completeness theorem for SCI. In fact, if non $X \vdash \alpha$ then, by the regularity of SCI, there exists a complete theory Z over X such that α is not in it. Then, the Lindenbaum-Tarski (quotient) model M_Z with respect to Z has the desired property: some valuation, viz., the canonical one over M_Z satisfies all formulas in X but does not satisfy α . Now we may build a c-model $C(M_Z)$. In view of (13.10), it has the same property.[LIII] So, one clearly sees that (13.4) is implicitly based on our semantics using SCI-models. The theorem (13.4) holds for special c-models because their kernels do the job.

Special c-models are sophisticated superstructures built ad hoc over SCI-models. One may seriously ask why build a ten foot computerized washer to wash just a single handkerchief? I do not know. Other observations are more definite, however. Given a c-model C, let $TR(C)$ be the set of all formulas true in C, i.e., satisfied by all homomorphism.s By (13.10), for every SCI-model M the set $TR(C(M))$ is a subset of $TR(M)$ but the converse relation does not hold, in general. Compare, for example, $C(M)$ with M if M is finite or, in particular, the Fregean model and you will clearly see that in possible worlds many quite impossible things happen. The lesson is that you cannot expect any smooth relationship between c-models and their SCI-theories. The c-model semantics does not seem to be very fruitful. Thus, we see that a completeness theorem for a logic is not a golden crown on a semantics but just the beginning of it. Already the relativity of completeness theorems forces us to ask how far can we go in a semanatics beyond(!) the completeness theorem. To construct a logical calculus and prove a kind of completeness theorem for it (not to speak about weak completeness theorem) appears very often just wasting time.

Possible-world semantics presents an even more acute case. Since it uses only powersets 2^I as universes the sentential variables run over, no possible-worlds model can be countably infinite. I see this as the strongest argument against this kind of semantics. Possible-worlds semantics cannot provide any theorem about countable infinite models as (6.4) or (9.19) because they do not exist at all. One need not to be a deep minded philosopher to know that the difference between countability and uncountability is as nearly important as that between finiteness and infinity. So, the possible-world semantics appears to be a real problem killer since it does not allow one to ask important questions concerned with countable infinite models.

14. Final remarks.

The fight with intensional ghosts is boring, unprofitable and, may never end. But, I must finish this paper. So, I conclude with some unelaborated remarks.

(1) Modal logicians think that NFL is a kind of modal logic and one badly done at that. Extensionally minded logicians consider FL as the only extensional logic. Thus, either intensional ghosts or the Fregean axiom and, everything else comes from the devil. I know this well. But, why is this so? This is a longer story. Indeed, you must first look at the history of modern logic and you will find there many relevant facts. Secondly, you must dig deeper and deeper in the non-Fregean logic and its relationship to the Fregean one.

(2) But, if you do that then you will certainly meet a fact which suggests that we may easily get rid of the identity connective and NFL. It is sufficient to translate NFL into FL as shown in [49] by Cresswell, again. Now, this is something quite different from

diffuse intensional ghosts. But things never are so simple as they seem. You know that translations may be better or worse and, the bad translations may not be translations at all. What always matters about translations is what are their invariants. If you ask for invariants of Cresswell's translation you will find only a few. It may be considered as a translation only if one wants to get rid of identity connective at any price. But, the real paradox is that it is impossible to get rid of the identity connective at all. What can be done is to equate it with material equivalence. Also, one cannot get rid of situations unless one agrees that thought is about nothing, or, rather, stops talking with sentences.

(3) Stimulated by Cresswell's "translation" and some ad hoc construction by Slupecki I built a theory of <u>reification</u> of <u>situations</u> in [50]. It resembles, in a sense, what Quine called reification of universals and also the relationship between Fregean FUNKTIONEN and their WERTVERLAUFE. Reification of situations is performable within a theory in some W-language, based on non-Fregean logic, of course. If based on a logic as strong as $\vdash_{WB}$ it automatically leads to what we are used to talking about in probability theory, viz., events being objects constituting a Boolean algebra in ordinary sense. Finally, this theory is the basis for a translation of Cresswell's type. This translation does not preserve entailment perfectly. But, equations are translatable exceptionally well, naturally.

(4) Non-Fregean logic contains the exact theory of facts, i.e., situations described in true sentences or, in other words, situations which obtain. If one accepts the Fregean axiom then one is compelled to be an absolute monist in the sense that there exists only one and necessary fact. So, you see that Fregean axiom should be abolished. Also, only the non-Fregean logic allows us to repeat with full understanding Wittgenstein's thesis that the real world is a totality of facts and not objects. Indeed, what I tried to do in [51] was a formal

reconstruction of the ontology of TRACTATUS, OT for short, as interpreted by Dr. B. Wolniewicz.[LIV] At that time, I only had a general idea of which logic OT has to be based on. Now, I may say that NFL suits OT and is the general and weakest extensional and two-valued logic.

(5) For many years I have viewed logic as an entailment. This point of view is responsible for the way the non-Fregean logic is constructed, presented and investigated. However, I cannot pretend to be the sole inventor of NFL and identity connective, in particular. This connective has a long history: J. Łos (1948), H. Greniewski (1957) and, A. N. Prior and M. J. Cresswell (1960^s). But, all those logicians except Łos, worshipped, more or less, intensional ghosts. This is evidently seen from [53] where the formulas $p \not\equiv \neg\neg p$, $p \not\equiv (p \wedge p)$ have been suggested as tautologies of SCI. Moreover, a methodology according to which logics are sets of theorems of formal systems prevented them to see and investigate NFL as I prefer to do. Yet, I would never know NFL as I know it now without my friendly young collaborators, Stephen L. Bloom and Donald J. Brown. It is essentially due to them that today NFL may be seen distinctly as a general logic with exceptionally good properties.

(6) Yet, this is not all. SCI-models are structures which may be considered as models of the Fregean logic which makes sentential variables trivially superflous. Thus, again, we may get rid of sentential variables and identity connective similarly as in the Cresswell translation. We may and we really do. This is the point. Of course, this fact does not discredit NFL at all but it hints to some tendency in our thinking which is codified, in a sense, by the Fregean axiom. Be it right or wrong, this tendency is certainly very strong and much older than modern logic. So it seriously calls for an explanation. This may not be easy. Why do we tend to describe the world as consisting of a single necessary fact? But, anyway, I feel

that when following this tendency we lose something. This is also suggested by the fact that Cresswell's translation is not so good. Moreover, one may say even that the conventional semantics which uses SCI-models is not adequate. Elements of these models are objects; but sentential variables do not actually run over any totality of objects. One cannot denote a situation by a name. It can only be described by a sentence. Thus, an adequate semantics of non-Fregean theories should use a W-language with non-Fregean entailment in it. Hence, we face the task of using(!) a new kind of language with a new entailment. I have shown to you how good this new logic is. Therefore, I think we should try to use it and I hope we will be successful. Frankly, however, I am not sure if we are able to do that.

Stevens Institute of Technology

August 1972.

S u p p l e m e n t s

I. The Fregean scheme applies not only to names and sentences but also to formators of diverse categories, e.g., connectives, functors, predicates, quantifiers and unifier (description formator). Accordingly, the famous Fregean ontological dictum should read:

(*) everything is either a situation or an object or a function.

To be sure, neither Fregean functions nor their courses-of-values will be discussed here. Nevertheless, both the syntax of W-languages (section 2) and the structure of their conventional (set-theoretical) models (section 4 of this paper and [11], [13]) clearly suggest that there is some significant truth in (*).

II. One may think that we have entered the area of many-valued logic, since sentences are allowed to have more than two values. But then, one has to realize that this is the referential (ontological) many-valuedeness, quite distinct from logical many-valuedeness. The latter is, probably, beyong my comprehension.

III. Elaboration of the Fregean scheme shows that the assignments $\underline{\underline{r}}$ and $\underline{\underline{t}}$ are of equal status as far as the construction of extensional logic is concerned. This suggests that the Fregean scheme conforms to the correspondence theory of truth. The assignment $\underline{\underline{s}}$ is, however, a secondary one as mentioned previously. You are advised not to jump to conclusions before the logic is constructed. Otherwise, you might be staggered by the idea of the eternal realm of senses (and propositions, in particular), independent of any language or logic and, prior to all referents. Subsequently, you would certainly find yourself in some impossible world. It must be emphasized here that one can prove (1.3), granted a two-valued extensional logic in a language with identity formator $\equiv$. One may define a kind of synonymity relation $\widetilde{T}$, relativized to a given (consistent) theory T: if both φ, ψ are terms or formulas then $\varphi \underset{T}{\sim} \psi$ means that the equation $\varphi \equiv \psi$ is in T. Then, $\underset{T}{\sim}$ is a congruence of the language, considered as free algebraic structure. Moreover, $\underset{T}{\sim}$ is the greatest congruence in the class of all congruences $\sim$ of the language which satisfy the condition: if $\varphi \sim \psi$ where φ, ψ are formulas and, φ is in T then ψ also is in T. Set $s_T(\varphi)$ = the abstraction class of φ modulo $\underset{T}{\sim}$. Then, T-senses of terms and formulas constitute an algebraic structure, similar to the language and known as the Lindenbaum-Tarski canonical structure (the quotient of the language) with respect to T. On the other hand, any model M defines a reference assignment r_M and, each model M of T induces a complete theory T_M over T; T_M = the theory of M. Then, $r_M(\varphi) = r_M(\psi)$ iff $\varphi \underset{T_M}{\sim} \psi$, i.e., $(\varphi \equiv \psi)$ is in T_M. Moreover, we can find a morphism <u>f</u> from the structure of T-senses <u>onto</u> the structure of M-referents, that is, <u>into</u> the model M such that in general: $\underline{f}(s_T(\varphi)) = r_M(\varphi)$. Compare [6], p. 21.

IV. In view of this, the basic laws of entailment formulated by Scott must be slightly modified and the reader can easily do that. Scott only considers finite entailment relations and, therefore, his antecedents of $\vdash$ are supposed to be finite sets of formulas. On the other hand, he allows the succedents of $\vdash$ to be arbitrary finite sets of formulas. Of course, finite entailment relations are of greatest interest. We drop the finiteness condition since, in the case of NFL, we wish to obtain it as a corollary. Also, the second difference is not very deep. The conditional assertion relation of Scott involves multiple sequents (with many succedents) since he follows G. Gentzen closely. This approach seems, however, to be useful only to prove, if true, the so called cut elimination theorem. Moreover, if Scott's entailment relation deals with a genuine disjunction the multiple sequents reduce to simple sequents (with exactly one succedent); compare the rule ($\vee$) in [1], p. 798. For the theory of entailment relations adopted here and its historical aspects the reader may consult [7] and [8]. Entailment relations $\vdash$ can always be replaced by the corresponding consequence operation Cn on sets of formulas. They are connected as follows:

$$\alpha \text{ is in } Cn(X) \text{ iff } X \vdash \alpha .$$

V. It may happen, for some pathological entailment $\vdash$ that $\vdash$ has no tautologies but, curiously enough, there exist formulas β such that $\alpha \vdash \beta$ for every formula α. Notice also that distinct(!) entailments may have a common set of tautologies(!).

VI. Languages of kind W may involve even many sorts of nominal variables and terms (names).

VII. Equations $\varphi \equiv \psi$ belong to W-languages under discussion, which is carried on in the meta-language. The meta-language also involves the notion of identity and equations which we render rather differently. There, we use the conventional symbol of equality. The distinction between $\equiv$ and $=$ only emphasizes the distinction between language and meta-language.

VIII. The collection of all truth-valuations is a Boolean space. Apply a compactness argument as in [8].

IX. If the language involves bound variables (quantifiers and unifier) then the natural postulate should be slightly modified.

X. Historical motivation for this terminology may be found in [18] and [19], p. 492. Henle algebras may be seen as Boolean algebras with an additional unary operation I such that

(*) $Ia = 1$ if $a = 1$ and, $Ia = 0$ if $a \neq 1$.

Conditions (*) and (4.9) are equivalent if we agree that:

(**) $a \circ b = I(a \dotplus b)$ and $Ia = a \circ 1$.

Henle algebras are algebraically simple and present an extreme in the class TBA of topological Boolean algebras, that is, Boolean algebras supplemented with interior operation I such that:

$$(*_*^*) \qquad I1 = 1 \qquad I(a \wedge b) = Ia \wedge Ib$$
$$Ia \leq a \qquad IIa = Ia$$

(compare [20]). When defining the class TBA one may also use a binary circle operation instead of I. They are mutually definable by (**); see [9], pp. 23-24 and, [21]. It has been shown in [21] that mutual definability (**) is also valid for the much broader class QTBA of quasi-topological Boolean algebras (Lemmon's extension algebras [23]). QTBA arises from TBA if we drop the last equation in $(*_*^*)$.

XI. Let $\underline{\underline{A}}$ be in TBA; f-sets exist in $\underline{\underline{A}}$ iff $\underline{\underline{A}}$ is well-connected in the sense of [22], i.e., equivalently,

(*) if $(a \circ b) \cup (c \circ d) = 1$ then $a \circ b = 1$ or $c \circ d = 1$.

If we pass to QTBA or, moreover, arbitrary Boolean algebras (with circle operation o) then the condition (*) for the existence of f-sets becomes more intricate.

XII. The condition (4.7) can be relaxed. Write $a \underset{F}{=} b$ whenever $a \circ b$ is in F and, replace (4.7) by:

(4.7.1) $\underset{F}{\equiv}$ is a congruence of $\underline{\underline{A}}$,

(4.7.2) if $a \underset{F}{=} b$ and a is in F then also b is in F.

Then, we get a much broader class of generalized models. Models defined by (4.7) deserve to be called normal. Generalized models can be contracted to normal one: factor by $\underset{F}{\equiv}$. If $\underline{\underline{M}}^*$ is the contraction of a generalized model $\underline{\underline{M}}$ then $\underline{\underline{M}}$ and $\underline{\underline{M}}^*$ have the same properties with respect to satisfaction, truth and entailment. The contraction $\underline{\underline{M}}^*$ of any normal model $\underline{\underline{M}}$ is isomorphic to $\underline{\underline{M}}$. This means that normal models are simple algebraic structures. It also means, if put in semantical terms, that normal models provide the exact interpretation of the identity connective.

XIII. It is easily seen that in the Fregean model the operation o and $\div$ are the same. This property is characteristic for the Fregean model. One may also consider generalized models in which o and $\div$ are equal and, call them generalized Fregean models. They can be very large even if the underlying algebra is Boolean. Generalised Fregean models, based on Boolean algebras, occur in structures known as Boolean models (for the Fregean logic, FL); see [20] and [24].

XIV. This is not to say that in every (normal) model the f-set contains at least two elements. There exist models $\langle \underline{\underline{A}}, F \rangle$, finite or infinite, where F has exactly one element. Then obviously, $\underline{\underline{A}}$ cannot be a Boolean algebra unless the complement A - F also has exactly one element. To be sure, models of this kind are intuitively very strange. But note that the great diversity of models reveals simply the real weakness of the SCI (and NFL).

XV. One may extend $\underline{r}$ even more. Using the convention concerning & and * one may set $\underline{r}$(&) = * and see that given a valuation $\underline{r}$ over $\underline{\underline{M}}$, everything (relevant) in the language has a referent in $\underline{\underline{M}}$ under $\underline{r}$. Sometimes, it appears convenient to consider pairs $\langle \underline{M}, \underline{r} \rangle$ where $\underline{M}$ is a model and $\underline{r}$ a valuation over $\underline{M}$. They are also called models, sometimes. In fact, we were concerned in III with this kind of model.

XVI. Consider the formulas γ_n for $n = 1,2,\ldots$:

$$(p_1 \not\equiv p_2) \wedge \ldots \wedge (p_i \not\equiv p_j) \wedge \ldots \wedge (p_n \not\equiv p_{n+1})$$

where p_i, p_j are variables and $1 \leq i < j \leq n+1$. Clearly, γ_n is satisfiable in M iff the universe of M contains at least n+1 elements.

XVII. In principle, every entailment can be defined that way, [8]

XVIII. For example, TFA = the set of all formulas of the form:

(1) $\alpha \Rightarrow (\beta \Rightarrow \alpha)$

(2) $(\alpha \Rightarrow (\beta \Rightarrow \gamma)) \Rightarrow ((\alpha \Rightarrow \beta) \Rightarrow (\alpha \Rightarrow \gamma))$

(3) $\neg\alpha \Rightarrow (\alpha \Rightarrow \beta)$

(4) $(\alpha \Rightarrow \beta) \Rightarrow ((\neg\alpha \Rightarrow \beta) \Rightarrow \beta)$

(5) $(\alpha \Leftrightarrow \beta) \Rightarrow (\alpha \Rightarrow \beta)$, $(\alpha \Leftrightarrow \beta) \Rightarrow (\beta \Rightarrow \alpha)$

(6) $(\alpha \Rightarrow \beta) \Rightarrow ((\beta \Rightarrow \alpha) \Rightarrow (\alpha \Leftrightarrow \beta))$

(7) $(\alpha \wedge \beta) \Leftrightarrow \neg(\alpha \Rightarrow \neg\beta)$, $(\alpha \vee \beta) \Leftrightarrow (\neg\alpha \Rightarrow \beta)$

If we modify (4) and (7) to suit intuitionistic requirements and, retained axioms for identity connective we would define a weakened SCI which might have its own, possibly interesting, model theory not necessarily based on possible-worlds semantics.

XX. Consider all equations as additional variables and directly apply the well known argument by L. Kalmer.

XIX. By schemes of inference or formulas we mean the sequential rules of [8].

XXI. This is a fundamental requirement of syntactic definition of NFL in comprehensive languages. For each formator, binding variables or not, a corresponding invariance axiom (scheme) with respect to identity (connective and/or predicate) must be laid down.

XXII. One might use either (5.8) instead of the invariance axioms (5.2), (5.2), (5.4) or (5.9) instead of the invariance axioms and the special identity axiom.

XXIII. The entailment $\vdash$ is <u>negative</u> ($X \vdash \alpha$ iff X and $\neg\alpha$ form together an inconsistent set). Hence, finiteness of $\vdash$ is equivalent to ℓ-compactness of $\vdash$ (every inconsistent set contains a finite inconsistent set). This implies the Lindenbaum theorem on complete extensions of consistent sets and regularity follows (again by negativity). Compare [25].

XXIV. An inductive argument, using length of derivations, shows that: if $X \vdash \alpha$ then $X \vdash^{(\underline{\underline{t}})} \alpha$. To show that: if $X \vdash^{(\underline{\underline{t}})} \alpha$ then $X \vdash^{(\underline{\underline{r}})} \alpha$, suppose non $X \vdash^{(\underline{\underline{r}})} \alpha$. There exists $\underline{\underline{M}}$ and $\underline{r}$ such that α is not satisfied but every formula in X is. Observe that the set Y of all formulas satisfied by $\underline{r}$ in $\underline{\underline{M}}$ is a complete theory. Put $\underline{t}(\beta) = 1$ if β is in Y and $\underline{t}(\beta) = 0$, otherwise. Clearly, non $X \vdash_{\underline{t}} \alpha$ and, hence, non $X \vdash^{(\underline{\underline{t}})} \alpha$. To show that: if $X \vdash^{(\underline{\underline{r}})} \alpha$ then $X \vdash \alpha$ we argue as follows. If non $X \vdash \alpha$ then, by regularity, X is contained in a complete theory Y. Factor both the language and Y by $\underset{Y}{\sim}$ (see III) to get the Lindenbaum-Tarski quotient model $\underline{\underline{M}}(Y)$ with respect to Y. Then, the canonical morphism $\underline{r}$, sending β into the abstraction class of β, is a valuation over $\underline{\underline{M}}(Y)$ such that Y is precisely the set of all formulas satisfied by $\underline{r}$ in $\underline{\underline{M}}(Y)$. Hence, non $X \vdash_{\underline{\underline{M}}(Y)} \alpha$ and, therefore, non $X \vdash^{(\underline{\underline{r}})} \alpha$.

XXV. Completeness theorems are seen here as suitable equations. Therefore, the often made distinction between soundness and proper completeness is overlooked. It only stresses the difference in the nature of proofs of these parts. Soundness is usually provable by induction. Proper completeness necessarily calls for a construction and is, as a rule, much harder to prove.

XXVI. Mostly, other equivalent formulations are used: every consistent set of formulas has a model or, is verifiable (and conversely).

XXVII. Weak completeness of a logic does not say much. Słupecki's "widerspruchsvertragendes Aussagenkalkul" [26] and Hiz's "complete sentantial calculus admitting extensions" [27] are examples of weakly complete truth-functional logics. The strangeness of these constructions can be explained only if we ask for the entailments involved. Indeed, they are not complete in our sense; compare [28], pp. 203-204.

XXVIII. Here, the role of complete theories is again crucial. First, we get all (up to isomorphism, countable) models as Lindenbaum-Tarski quotient models modulo complete theories. Secondly, complete theories correspond (one-one) to truth valuations.

XXIX. We assume structurality also, in the sense of [8], which is almost no restriction at all.

XXX. We say that $\vdash_1$ is weaker (or equal) than $\vdash_2$ and, $\vdash_2$ is an extension of $\vdash_1$ iff $X \vdash_2 \alpha$ whenever $X \vdash_1 \alpha$. Then, clearly, every $\vdash_2$ theory is a $\vdash_1$ theory but not necessarily conversely.

XXXII. The fact that $\vdash_G$ is not an elementary extension of means that the G-rule cannot be, in general, replaced by any additional set of axioms. On the other hand, each particular G-theory T obviously equals to TH(X) for some X. One may ask for a nice and, perhaps, independent X such that TH(X) = the smallest G-theory.

XXXI. In 1969, a pupil of the late R. Montague wrote 36 pages to show that non-Fregean logic is rooted in the logic of modality. He tried to prove the Fregean axiom and convince me that instead of NFL and, in particular, SCI something else should be constructed. Naturally, if you do not want NFL then you will get something else.

XXXIII. Let $\vdash^*$ be the entailment defined syntactically by LA, MP and the substitution rule Sb. Clearly, $\vdash^*$ is a non-elementary extension of $\vdash$ but the rule Sb has exceptionally nice properties. Since, Sb(TH(X)) is contained in TH(Sb(X)) we infer that the invariant theories constitute precisely the collection of theories of $\vdash^*$ and, a modification of Wojcicki's method works perfectly with $\vdash^*$; [30], [25]. The semantics of $\vdash^*$ is a special case of that for $\vdash$ and offers some interesting facts. The essential point is that instead of $\vdash_M$ we must use the entailment $\vdash^*_M$, defined for each model M, as follows: $X \vdash^*_M \alpha$ iff α is in TR(M) whenever the whole set X is contained in TR(M).

XXXIV. If M is adequate for $\vdash_T$ then, obviously, M is adequate for T. Conversely, if M is adequate for T then M is adequate for $\vdash_T$ iff $\vdash_M$ is finite (1-compact). But, Stephen L. Bloom showed that if M is adequate for T then some ultrapower of M is adequate for $\vdash_T$. The ultraproduct construction of Łos and his main theorem fit perfectly into the semantics of NFL and SCI, in particular. One may dare to say that only the theory of models of NFL reveals the real nature and value of the ultraproduct construction.

XXXV. NFL in open W-languages also provides other cases of that analogy. Łos theorem on regularity of $\vdash^*$ with respect to the collection of quasi-complete theories holds in open W-languages; see [13]. Moreover, the theorem on common extensions of models (like in

[34]) is also valid for models of open W-languages (unpublished).

XXXVI. If M is any finite model such that the algebra of it has no proper subalgebras then the theory TR(M) is Post-complete (S. L. Bloom).

XXXVII. Theorem (9.8) is concerned with those (ordinary) Boolean algebras with circle operation which involve f-sets (ultrafilters with property (4.7)). The class of these algebras may be axiomatized as follows. First, write axioms corresponding to (9.1), (9.2) and (9.3). They are equations with identity predicate:
$(x \cap y) \cup z \equiv (y \cup z) \cap (x \cup z)$, $(x \cup y) \cap z \equiv (y \cap z) \cup (x \cap z)$, $x \cup (y \cap -y) \equiv x$, $x \cap (y \cup -y) \equiv x$, $x \dot{-} y \equiv -x \cup y$ and $x \div y \equiv (x \dot{-} y) \cap$ $\cap (y \dot{-} x)$. Subsequently, let $x \leq y$ stand for $x \dot{-} y \equiv 1$ and, add the following infinite list of implications (n,m = 1,2, ...):

(*) if $(z_1 \circ z_1) \cap \ldots \cap (z_n \circ z_n) \leq (x_1 \circ y_1) \cup \ldots \cup (x_m \circ y_m)$
then either $x_1 \equiv y_1$ or ... or $x_m \equiv y_m$.

Finally, add $x \not\equiv -x$. If you compare this infinite axiom set with axioms of WB then you have a case of difference, again.

XXXVIII. To see here a case of analogy, recall XI and note the following semantical theorem: if $M = \langle A, F \rangle$ is a model then M is a model of WT iff A is a (well-connected) topological Boolean algebra.

XXXIX. To point out the difference between WB and WT, the following formal facts may prove to be of some interest. The Boolean law $[(p \wedge q) \equiv p] \Leftrightarrow [(p \vee q) \equiv q]$ is in WB. However, the equation $[(p \wedge q) \equiv p] \equiv [(p \vee q) \equiv q]$ is in WT but not in WB. Compare also the second formula in (9.7) and the first one in (9.13). Thus, there is a certain ambiguity in WB with respect to the ordering of the

Boolean algebra of situations. On the other hand, let A (or, B) be the set of all equations $\alpha \equiv \beta$ of the SCI-language such that the equivalence $\alpha \Leftrightarrow \beta$ is in TFT (or, TAUT). Then WB = TH(A) and WT = TH(B).

XL. Instead of (9.16), one may also use (together with $_T$) any one of the following formulas: $(p \equiv \Box p) \Leftrightarrow (\neg p \equiv \Box \neg p)$, $(P \not\equiv q) \equiv [(p \equiv q) \equiv O)]$, $(p \equiv \Box p) \Rightarrow [(p \equiv 1) \vee (p \equiv O)]$.

XLI. As you might expect, equational definitions have all the regular syntactic properties (translatability, non-creativity and eliminability) with respect to the absolute non-Fregean entailment in any W-language. But given a W-language with quantifiers and equations of both kind, one may ask for properties of so called standard definitions of the form

$$u \equiv \Phi(v_1, \ldots, v_n) \Leftrightarrow \alpha[v_1, \ldots, v_n, u]$$

where Φ is a new formator and we know that for all $v_1, \ldots, v_n$ there exists exactly one u such that $\alpha[v_1, \ldots, v_n, u]$. It is easy to prove that standard definitions actually define semantically new functions in models. However, we cannot prove that standard definitions have all the regular syntactic properties unless we supplement the entailment with suitable ontological principles. The details of the problem of definitions in non-Fregean logic are being now worked out in collaboration with Mr. M. Omyla, a doctoral candidate at Warsaw University. As yet, we know, for example, that a nice theory of standard definitions requires Boolean assumptions (like in WB) and also a Q-principle which states, roughly speaking, that the quantifiers $\forall$ and $\exists$ are signs of generalized meets and joins in the Boolean algebra of situations.

XLII. For the proof, observe first that $\langle A,F \rangle$ is a generalized model of WT whenever A is any algebra in TBA and F is any ultrafilter of A. Use then the contraction operation and the well-known McKinsey-Tarski characterisation of S_4 in [38]. Notice, in connection with this, that McKinsey found in [39] a generalised model of hypercontinuum power, adequate for WT. Infinite models adequate for WH (countable or uncountable) are due to Scroggs [40].

XLIII. But suppose, $\vdash^m$ is the modal entailment. If you want to compare $\vdash^m$ with $\vdash$ or some extension of it you have to use another technique of R. Wojcicki, again. His note [41] will also show you how little information on relation of entailments is provided by translation maps between corresponding sets of tautologies.

XLIV. There is an invariant equational Boolean theory, properly contained in WT and very close to Feys-von Wright system S_o. Put $WT^* = TH(Sb(\alpha_T^*))$ where α_T^* is like α_T with $(\Box 1) \equiv 1$ in place of $\Box\Box p \equiv \Box p$. WH* is not a G-theory. It contains $\Box\Box p \Leftrightarrow \Box p$ and $(p \equiv q) \Leftrightarrow \Box(p \equiv q)$. These formulas are not in S_o, obviously. But, believe it or not, WT* and S_o contain exactly the same equations if S_o is considered as a part of WT*. Moreover, there is another relationship between S_o and WT*. By Lemmon's result [23] the system S_o may be thought as a "theory" of algebras in QTB or, more precisely, quasi-topological Boolean algebras with the unit as only one distinguished element. On the other hand, WT* is the theory (of SCI) of exactly those models $\langle A,F \rangle$ where A is a quasi-topological Boolean algebra. We conclude that Lemmon's algebraic semantics does not fit extensional two-valued logic. Why not? You may think about that. Clearly, you will meet first the difference between the unit of a Boolean algebra and ultrafilters in it such that (4.7) holds. Thereby, you will taste some whim of intensionality. But, I am not interested in it. Even more, if you are really interested in non-Fregean logic then I advise you to forget all modal systems which are not SCI-theories. Also, you better forget the labels S_4, S_5 and the like. Why start counting with 4?

XLV. Modal logic is an established trend in formal logic of today. It starts with intensional notions of necessity, possibility and the like and, constructs formal systems which alledgedly formalize these nitions but involve many perplexities, mostly concerned with identity. Eventually, it lives in the mad philosophy of intensional entities, non-existent objects and essentialism. Clearly, any endeavour to make NFL subordinate to modal logic is simply an arrogant offence.

You may be sure that the fountain of modal logic will never dry up. However, it is only a kind of naivete to attempt to build an intensional formal logic. Whatever intensional formal logic you might propose then, sooner or later, I bet someone will invent an intensional construction in natural language which breaks all your logical rules. And no harm will be done to our discourse. This, on the other hand, cannot happen with NFL. If one starts with true premisses and breaks the laws of the non-Fregean logic then one will arrive, sooner or later, at false conclusions.

XLVI. In open W-languages, the set LA of logical axioms is to be augmented with axioms for identity predicate: $\tau \equiv \tau$ and invariance axioms. Note that we must also have the "mixed" invariance axiom scheme:

$$(\tau_1 \equiv \eta_1) \wedge (\tau_2 \equiv \eta_2) \Rightarrow [(\tau_1 \equiv \tau_2) \equiv (\eta_1 \equiv \eta_2)]$$

where $\tau_1, \tau_2, \eta_1, \eta_2$ are any terms (nominal formulas). Furthermore, the theory WB is defined as in SCI. However, we define WT as in XXXIX. To get WH we add the principle: $((\varphi \equiv \psi) \equiv 1 \vee (\varphi \equiv \psi) \equiv 0)$ where φ, ψ are any formulas, both sentential or both nominal.

XLVII. We may suppose the underlying language to be open. If, however, the language involves quantifiers then (11.15) requires a stronger assumption on T.

XLVIII. Possible-world semantics for modal systems with quantifiers forces us to consider fiction stories in which, for example, Sherlock Holmes was a scholastic logician at Oxford and Dr. Watson invented possible-world semantics. Some tales of this kind may even be amusing. Yet, this is not a science and logicians are not necessarily the best fiction writers.

XLIX. The semi-Fregean postulate seems to be a mixture of a mathematical desire for simplicity and some intensional whim. The Boolean unit is distingished by properties of Boolean operations only. Thus, a logical matrix based on a Boolean algebra and containing the singleton $\{1\}$ as the set of distinguished elements may be replaced naturally by the underlying Boolean algebra. In this case we get rid of logical matrices in favour of mere algebras and, a superficial simplification actually is attained. However, this is not a complete description of the semi-Fregean postulate. For example, Kripke models constitute a rather complex kind of relational structure with the power 2^I as the underlying universe; also, the unit of the Boolean algebra 2^I is used in modal logic as the only distinguished element.

L. Although I am strongly opposed to the intensional formal logic, non-existent entities and propositions independent of any language and, at the same time, to the Fregean axiom in extensional logic I do not thereby attack the general index method in semantics. It may be considered as an endeavour to "dilute" the semantics of unambigous extensional logic to describe ambiguities and vagueness of our thinking in natural language (compare [47], [48]). Unfortunately, the index method is closely tied to intensional logic. Possible-worlds semantics in modal logic is the first application of index method when the

Fregean logic was a paradigm of extensional and two-valued logic. What I can say right now is only methodological advice: develop first the non-Fregean logic as a more general paradigm of extensional and two-valued logic and, only then try to use the index method.

LI. The original definition of Cresswell models and the concept of satisfaction is formulated within the customary formalism of possible-worlds semantics. One may easily transform Cresswell models into c-models using one-one correspondences between relations in $I \times P(I) \times \ldots \times P(I)$ and maps from $P(I) \times \ldots \times P(I)$ to $P(I)$ (where $P(I)$ is the set of all subsets of I) and, between $P(I)$ and 2^I. The final form of c-models allows us to define satisfaction by means of homorphisms as in case of our models. A homomorphism from the SCI-language to a c-algebra G is a map h of all formulas into 2^I such that for all α, β : $h(\neg\alpha) = G_1(h(\alpha))$, $h(\alpha \wedge \beta) = G_2(h(\alpha), h(\beta))$ and $h(\alpha \equiv \beta) = G_0(h(\alpha), h(\beta))$.

LII. One may begin with a generalized SCI-model M and obtain, in a quite analogous way c-models $C(M)$ which also satisfy (13.10). Moreover, one may contract each such $C(M)$ to a structure which is isomorphic to some c-model $C(M^*)$ where M^* is the contraction of M.

LIII. Actually, the Cresswell proof uses the generalized SCI-model N_Z which consists of the SCI-language and the theory Z and, subsequently, the special c-model $C(N_Z)$.

LIV. However, Herr Gruppenfuhrer der SL (i.e., Symbolic Logic) Standarte, Georgie Top, disagrees. Actually, it does not matter here, at all. Another story is that the subtle logician, John Myhill, suggested that OT should be rather formulated within the set theory (based on FL, of course). Evidently, the reification of situations is at work in Myhill's mind. In fact, he sketched a conventional (set-theoretical) model of OT. But, one obtains the consistency of OT and not the OT itself, that way. Consistency of OT has been proved in [52] in a geometrical way (using the Cantor set).

BIBLIOGRAPHY

1) Scott, D: On engendering an illusion of understanding, The Journal of Philosophy 68(1971), 787-807.

2) Quine, W.V.: Comments, in Boston Studies in the Philosophy of Science. D. Reidel Publishing Co., Dordrecht 1963, pp. 97-104.

3) Scott, D: Advice on Modal Logic, in Philosophical Problems in Logic, edited by K. Lambert, D. Reidel Publ. Comp. 1970, pp. 143-173.

4) Weyl, H: The Ghost of Modality, in Philosophical Essays in memory of E. Husserl, edited by M. Farbes, Harvard Univ. Press, 1940, reprinted by Greenwood Press, New York, 1956.

5) Church, A: Introduction to mathematical logic, Vol. I, Princeton, N.J. 1968.

6) Suszko, R: An Essay in the formal theory of extension and of intension, Studia Logica 20(1967), 7-34.

7) Bernays, P: Betrachtungen zum Sequenzenkalkul, in contributions to Logic and Methodology in honor of I.M. Bochenski, edited by A.T. Tymieniecka in collaboration with Ch. Parsons, North Holland, Amsterdam 1965, pp. 1-44.

8) Los, J. & Suszko, R: Remarks on sentential logics, Proc. Neth. Akad. van Weten., Ser. A, 67(1958), 177-183.

9) Suszko, R: Identity connective and modality, Studia Logica 27 (1971), 7-39.

10) Suszko, R: Non-Fregean Logic and Theories, Analele Universitatii Bucuresti, Acta Logica 9(1968), 105-125.

11) Bloom, S.L.: A completeness theorem for "Theories of Kind W", Studia Logica 27(1971), 43-55.

12) Quine, W.V.: From a Logical Point of View, Harvard Univ. Press, Cambridge, Mass., 1953.

13) Suszko, R.: Quasi-Completeness in non-Fregean Logic, Studia Logica 29(1971), 7-14.

14) Bloom, S. & Suszko, R.: Semantics for the Sentential Calculus with Identity, Studia Logica 28(1971), 77-81.

15) ____________________: Investigations Into the Sentential Calculus with Identity, Notre Dame Journal of Formal Logic, 13 (1972), 289-308.

16) Suszko, R.: Sentential Calculus with Identity (SCI) and G-theories, an abstract to appear in The Journal of Symbolic Logic 36(1971).

17) __________: SCI and Modal Systems, an abstract to appear in The Journal of Symbolic Logic 37(1972).

18) Scroggs, S.J.: Extensions of the Lewis System S_5, The Journal of Symbolic Logic 16 (1951), 112-120

19) Lewis, C.I., Langford, C.H.: Symbolic Logic, 2nd ed., Dover Publications, 1959.

20) Rasiowa, H., Sikorski, R.: The Mathematics of the Metamathematics PWN Warsaw, 1963.

21) Kagan, J.D.: An Axiomatization of Topological Boolean Algebras to appear in Zeitschrift fur Mathematische Logik und Grandlagen der Mathematik.

22) McKinsey, J.C.C., Tarski, A.: The algebra of topology, Annals of Mathematics 45(1944), 141-191.

23) Lemmon, E.J.: An extension algebra and the modal system T, Notre Dame Journal of Formal Logic 1(1960), 3-12.

24) Scott, D.: A Proof of Independence of the Continuum Hypothesis, Mathematical Systems Theory 1(1967), 89-111.

25) Brown, D.J., Suszko, R.: Abstract Logics, Dissertationes Mathematicae, Vol. 102(1973).

26) Stupecki, J.: Uber die Regeln des Aussagenkalkuls, Studia Logica 1(1953), 19-40.

27) Hiz, H.: Complete sentential calculus admitting extensions, Summer Institute of Symbolic Logic in 1957 at Cornell University, Vol. 2, 260-262.

28) Suszko, R.: Concerning the method of logical schemes, the notion of logical calculus and the role of consequence relations, Studia Logica II(1961), 185-214.

29) Wojcicki, R.: Logical matrices adequate for structural sentential calculi, Bull., Acad. Polon. Sci. (Ser. Sci. Math. Astronom. Phys.) 17(1969), 333-335.

30) Bloom, S.L., Brown, D.J., Suszko, R.: Some theorems on abstract logics, Algebra i Logika Sem. g(1970), 274-280.

31) Suszko, R.: Adequate models for the non-Fregean sentential calculus (SCI), in Logic, Language and Probability, edited by R.J. Bogdan and I. Niiniluoto, D. Reidel, Pub., (1973),pp. 49-54.

32) Bloom, S.L., Brown, D.J.: Classical abstract logics, in Dissertationes Mathematicae, Vol. 102(1973).

33) Los, J.: The algebraic treatment of the methodology of elementary deductive systems, Studia Logica 2(1955), 151-211.

34) Los, J., Suszko, R.: On the extending of models II, Fundamenta Mathematicae 42(1955), 343-347.

35) Hallden, S.: On the semantic non-completeness of certain Lewis calculi, the Journal of Symbolic Logic 16(1951), 127-129.

36) Diamond, A.H.: Simplification of the Whitehead-Huntington set of postulates for the algebra of logic, Bulletin of the American Mathematical Society, Vol. 40(1934), 599-601.

37) Suszko, R., Zandarowska, W.: Lewis' systems S_4 and S_5 and the identity connective (in Polish), Studia Logica 29 (1971), 169-177.

38) McKinsey, J.C., Tarski, A.: Some theorems about the Lewis and Heyting calculi, The Journal of Symbolic Logic 13(1948), 1-15.

39) McKinsey, J.C., Systems of modal logic which are not unreasonable in the sense of Hallden, The Journal of Symbolic Logic 18(1953) 109-113.

40) Scroggs, S.J.: Extensions of the Lewis system S_5, The Journal of Symbolic Logic 16(1951), 112-120.

41) Wojcicki, R.: On reconstructability of classical propositional logic in intuitionistic logic, Bull. Acad. Polon. Sci. (Ser. Sci. Math. Astronom. Phys.) 18(1970), 421-422.

42) Kripke, S.A.: Semantical considerations on modal logic, Acta Philosophica Phennica 16(1963), 83-94.

43) Lemmon, E.J.: Algebraic semantics for modal logics, The Journal of Symbolic Logic 31(1966), 46-65 and 191-218.

44) Makinson, D.C.: A generalisation of the concept of a relational model for modal logic, Theoria 36(1970), 331-335.

45) Lemmon, E.J.: A note on Hallden-incompleteness, Notre Dame Journal of Formal Logic 7(1966), 296-300.

46) Cresswell, M.J.: Classical intensional logics, Theoria 36(1970), 347-372.

47) Montagne, R.: Pragmatics in R. Klibansky ed., contemporary philosophy-la philosophie contemporaine, Florence 1968, 102-122.

48) ____________: Universal grammer, Theoria 36, 1970, 373-398.

49) Cresswell, M.J.: Functions of propositions, The Journal of Symbolic Logic 31(1966), 545-560.

50) Suszko, R.: Reifikacja sytuacji (Reification of situations), Studia Filozoficzne 69(1971), 65-82.

51) __________: Ontology in the Tractatus of L. Wittgenstein, Notre Dame Journal of Formal Logic 9(1968), 7-33.

52) __________: Consistency of some non-Fregean theory, Notices of A.M.S. 16(1969), abstract 664-23.

53) Prior, A.N.: Is the concept of referential opacity really necessary? Acta Philosophica Fennica 16(1963), 189-198.

A REALIZABILITY INTERPRETATION OF THE THEORY OF SPECIES

William W. Tait

1. The main purpose of this paper is to define a realizability interpretation for second order intuitionistic logic and to prove that every deducible sentence of this system is realizable. As a consequence of this, we shall obtain Girard's 1970 Normalizability Theorem for second order intuitionistic logic. The ideas of this paper extend in a straightforward way to the intuitionistic simple type theory (logic of finite orders); but I will not discuss this extension here.

2. The Theory $\mathcal{S}$ of Species.

2.1. For each of the symbols α=i, 0, 1, ..., $\mathcal{S}$ contains infinitely many α-variables, denoted by

$$X^\alpha, Y^\alpha, Z^\alpha, X_0^\alpha, \dots .$$

The i-variables are intended to range over individuals and are also denoted by

$$x, y, z, x_0, \dots .$$

The 0-variables are intended to range over propositions and, for n>0, the n-variables over n-ary species of individuals (i.e. properties of n-tuples of individuals). The non logical constants of $\mathcal{S}$ are individual and function constants. An n-ary function constant (n>0) is intended to denote an n-ary operation on individuals. The individual or i-terms are built up in the usual way from i-variables and individual and function constants, and are denoted by

$$S^i, T^i, S_0^i, \dots, s, t, s_0, \dots .$$

The formulae of $\mathcal{S}$ are built up in the usual way from atomic formulae

$$X^n t_1 \dots t_n$$

n≥0, by means of the operations

$$A \supset B \qquad \forall X^\alpha A$$

for α=i, 0, 1, ... We shall see in 2. that the remaining intuitionistic logical operations can be defined in terms of these. If A is a formula, $n \geq 0$ and $x_1, \ldots x_n$ are distinct, then

$$T^n = \lambda x_1 \ldots x_n A$$

is called an n-ary predicate or n-term. We shall write

$$T^n t_1 \ldots t_n = A[{}^{t}1/x_1, \ldots, {}^{t}n/x_n]$$

where the right hand side denotes the usual substitution of i-terms for free occurrences of i-variables. Thus, $T^n t_1 \ldots t_n$ is a formula.

$$B[{}^{T^n} / X^n]$$

denotes the result of replacing each part $X^n t_1 \ldots t_n$ of B, such that the given occurrence of X^n is free in B, by $T^n t_1 \ldots t_n$. In all of these substitutions, we may have first to rename bound variables in order to avoid confusing them with free ones. We shall assume that this is done in some unique way.

Formal deductions in $\mathcal{S}$ are in the style of natural deduction (Gentzen 1934). They are finite trees which are built up from certain initial trees by means of rules of inference. The initial trees are the premises

$$\overset{n}{A}$$

consisting of a formula A with an index $n \geq 0$ over it. There are four rules of inference, an introduction rule and an elimination rule for each of $\supset$ and $\forall$.

$\supset$ I. $$\frac{\begin{matrix}\vdots\\ B\end{matrix}}{A \supset B}\, n$$

$\supset$ E. $$\frac{\begin{matrix}\vdots\\ A \supset B\end{matrix} \quad \begin{matrix}\vdots\\ A\end{matrix}}{B}$$

$\forall$I. $$\frac{\begin{matrix}\vdots\\ A\end{matrix}}{\forall X^{\alpha} A}$$

$\forall$E. $$\frac{\begin{matrix}\vdots\\ \forall X^{\alpha} A\end{matrix} \quad T^{\alpha}}{A[{}^{T^{\alpha}} / X^{\alpha}]}$$

Thus, in each of these rules, we take given deductions, eg of B or of $A \supset B$ and A, and combine them to obtain new deductions, eg of $A \supset B$ or of B, resp. We must make a restriction on $\forall$I; but first, we must say when an occurrence of a premise in a deduction is discharged: $\overset{n}{A}$ is undischarged in $\overset{n}{A}$. In each of $\supset$E, $\forall$I and E, an occurrence of a premise is discharged iff it is discharged in one of the given deductions. In $\supset$ I, an occurrence of a premise is discharged iff it is discharged in the given deduction of B or else the premise is A^n.

<u>Restriction on $\forall$I.</u> X^α must not occur free in any undischarged premise in the given deduction of A.

A is deducible in $\mathcal{S}$ (deducible in $\mathcal{S}$ from $B_0, \ldots, B_n$) iff there is a deduction in $\mathcal{S}$ ending with A and with no undischarged premises (all of whose undischarged premises are of the form B_i^k for some $i \leq n$).

2.2. Prawitz 1968 showed that the logical constants $\perp$ (absurdity), $\neg$, $\vee$, $\wedge$ and $\exists$ are definable in $\mathcal{S}$ by

$$\perp = \forall X^0 X$$

$$\neg A = A \supset \perp$$

$$A \vee B = \forall X^0((A \supset X) \supset ((B \supset X) \supset X))$$

$$A \wedge B = \forall X^0((A \supset (B \supset X)) \supset X)$$

$$\exists Z^\alpha B = \forall X^0(\forall Z^\alpha(B \supset X) \supset X)$$

With these definitions, all of the intuitionistic laws of logic are deducible in $\mathcal{S}$. Set

$$A \equiv B = (A \supset B) \wedge (B \supset A)$$

$$E = \lambda xy\ Z^1(Zx \equiv Zy)$$

Then the theory of identity of individuals is deducible in $\mathcal{S}$ in terms of E.

Assume that $\mathcal{S}$ contains the individual constant o and the unary function constant '. Set

$$N = \lambda xy \forall Z^1[(Z0 \wedge \forall y(Zy \supset Zy')) \supset Zx]$$

$$G_+ = \lambda xyz\ Z^3[(\forall u Zuou \wedge \forall uvw(Zuvw \supset Zuv'w')) \supset Zxyz]$$

$$G_\times = \lambda xyz\ Z^3[(\forall u Zuoo \wedge \forall uvwt(Zuvw \wedge G_+ wut \supset Zuv't) \supset Zxyz]$$

Then, à la Dedekind 1888, we can derive Peano's postulates for number theory from the premises

$$\forall xy(Exy \vee \neg Exy) \quad \forall x \neg Eox'$$

$$\forall xy(Ex'y' \supset Exy)$$

Let A^- be obtained from A by replacing each part $X^n t_1 \ldots t_n$ of A by $\neg\neg X^n t_1 \ldots t_n$. As noted by Kreisel, the result of Godel 1932-3 extends to show that if A is deducible in classical second order logic, then A^- is deducible in $\mathcal{S}$. (We are assuming here that A is a formula of $\mathcal{S}$.) One need only note that $\neg\neg A^- \equiv A^-$ is derivable in $\mathcal{S}$, that each of $\supset$I, $\supset$E and $\forall$I is preserved under the translation $A \to A^-$, and that $\forall$E translates as

$$\frac{\forall X^\alpha A^- \quad T^{\alpha-}}{A[T^\alpha / X^\alpha]^-}$$

When $\alpha = i$, this is again an instance of $\forall$E.
But, when $\alpha = n$, it is

$$\frac{\forall X^n A^- \quad T^{n-}}{A^-[T^{n-} \ / \ X^n]}$$

However, we do have the application

$$\frac{\forall X^n A^- \quad \neg\neg T^{n-}}{A^-[\neg\neg T^{n-} \ / \ X^n]}$$

of $\forall$E, from which we obtain $A[T^n / X^n]^-$ using $\neg\neg T^n t_1 \ldots t_n^- \equiv T^n t_1 \ldots t_n^-$.

Thus, in this sense, classical second order logic is embedded in $\mathcal{S}$. This result applies also to simple type theory.

2.3. It will be convenient and suggestive to introduce another notation for natural deductions in $\mathcal{S}$. Accordingly, we introduce the notion of an A-term for each formula A, by induction.

$$a \models A$$

means that a is an A-term. Let V_0, V_1, ... be an infinite list of symbols. X, Y, Z, X_0, etc. will denote arbitrary V_n.

$$X^A \models A .$$

X^A is called an A-variable.

$$b \models B \Rightarrow \lambda X^A b \models A \supset B$$

$$b \models B \Rightarrow \lambda X^\alpha b \models \forall X^\alpha B$$

$$c \models A \supset B,\ a \models A \Rightarrow (ca) \models B$$

$$c \models \forall X^\alpha B \Rightarrow (cT^\alpha) \models B[T^\alpha / X^\alpha]$$

The free occurrences of variables in X^A are X^A itself and the free occurrences of variables Y^α in A. The free occurrences in $\lambda X^A b$ and $\lambda X^\alpha b$ are those in b other than occurrences of X^A and X^α, resp. The free occurrences in (ca) and (cT^α) are those in c and those in a and T^α, resp. We must impose a

Restriction on λX^α. $\lambda X^\alpha b$ is a $\forall X^\alpha B$ term only when X^α is not free in A whenever Y^A is free in b.

With each natural deduction D of A we associate an A-term $|D|$ as follows:

$$\left|\begin{matrix} n \\ A \end{matrix}\right| = V^A_n$$

$$\left|\frac{\begin{matrix}\vdots \\ B\end{matrix}}{A \supset B}\, n\right| = \lambda V^A_n \left|\begin{matrix}\vdots \\ B\end{matrix}\right|$$

$$\left|\frac{\begin{matrix}\vdots \\ A\end{matrix}}{\forall X^\alpha A}\right| = \lambda X^\alpha \left|\begin{matrix}\vdots \\ A\end{matrix}\right|$$

$$\left|\frac{\begin{matrix}\vdots \\ A \supset B\end{matrix} \quad \begin{matrix}\vdots \\ A\end{matrix}}{B}\right| = \left(\left|\begin{matrix}\vdots \\ A \supset B\end{matrix}\right| \left|\begin{matrix}\vdots \\ A\end{matrix}\right|\right)$$

$$\left|\frac{\begin{matrix}\vdots \\ \forall X^\alpha A\end{matrix} \quad T^\alpha}{A[T^\alpha / X^\alpha]}\right| = \left(\left|\begin{matrix}\vdots \\ \forall X^\alpha A\end{matrix}\right| T^\alpha\right)$$

Note that discharged and undischarged premises transform into bound and free variables, resp. Hence, the restriction on $\forall$I transforms into the restriction on λX^{α}. Thus, $D \rightarrow |D|$ is a bi unique correspondence between natural deductions of A in $\mathcal{S}$ and A-terms. In view of this, we may use the language of terms from now on, rather than of natural deductions.

2.4. In formulating Gentzen's 1934 Hauptsatz for the system of natural deduction, Prawitz 1965 introduced conversion rules for deductions which, in the language of terms, are just Church's 1932 lambda conversion:

$$(\lambda x^A b)c \Rightarrow b[^c/x^A]$$

$$(\lambda X^{\alpha} b)T^{\alpha} \Rightarrow b[^{T^{\alpha}}/X^{\alpha}]$$

where the substitutions on the right for free occurrences of variables are defined in the usual way. From now on, we shall not distinguish between formulae or terms which can be obtained from one another by renaming bound variables.

$$a \gg b$$

will mean that b is obtained by replacing a part c of a by d, where $c \Rightarrow d$. If there is no b such that $a \gg b$, then a is called normal. A reduction is a sequence $a_0, a_1, \ldots$ (finite or infinite) such that $a_i \gg a_{i+1}$. a_0 is itself a reduction of length 1.

a b

means that there is a reduction beginning with a and ending with b. a is normalizable iff there is a normal b with $a \geq b$. a is well-founded (strongly normalizable in Prawitz 1970) iff all reductions beginning with a are finite. If a is well-founded, it is clearly normalizable.

<u>Well Foundedness Theorem for $\mathcal{S}$</u>. If $a \models A$, then a is well-founded.

Girard 1970, Martin-Lof 1970 and Prawitz 1970 all proved that every A-term is normalizable. (All of the proofs are based on an idea of Girard.) The arguments easily modify, however, to prove well-foundedness. We shall obtain the Well-Foundedness Theorem below as a consequence of the Realizability Theorem (whose proof is again not much more than a reformulation of Girard's idea).

The isomorphism between natural deduction and the lambda calculus with type structure, utilized in 2.3 - 4, was first noted in Curry-Feys 1958 for the case of positive implicational logic.

3. The Lambda Calculus and Realizability.

3.1. The terms of the lambda calculus $\mathcal{T}$ are built up from the variables V_n and the constant K by means of the operations

$$\lambda Xb \qquad (ab)$$

of lambda abstraction and evaluation, resp. The notions of a free occurrence of a variable, and of substitution for free variables are defined as usual. The rule of lambda conversion is

$$(\lambda Xb)a \qquad b[^a/x]$$

The relations $a \gg b$ and $a \geqslant b$ and the notions of normality, normalizability and well-foundedness are defined just as before in 2.4. Note that, unlike $\mathcal{S}$, $\mathcal{T}$ has non normalizable terms, eg $s = ((\lambda X(XX))\ (\lambda X(XX)))$; and also has normalizable non well-founded terms, eg $((\lambda XK)s)$. We shall write

$$ab = (ab),\ abc = ((ab)c),\ abcd = ((abc)d), \text{ etc.}$$

$$M$$

will denote the set of closed well-founded terms of $\mathcal{T}$. If $a \in M$ and $a \geqslant b$, then $b \in M$

$$a \equiv b$$

means that for some c, $a \geqslant c$ and $b \geqslant c$. The Church-Rosser Theorem (1936) asserts that $\equiv$ is an equivalence relation on the terms of $\mathcal{T}$.

We shall need the following result, which we prove in 4.

<u>Lemma</u> If $b[^a/x]c_1 \ldots c_n$ and a are in M, then so is $(\lambda xb)ac_1 \ldots c_n$ $(n \geqslant 0)$.

3.2. Let I^i denote the set of all closed i-terms of $\mathcal{S}$.

A <u>proposition</u> <u>over</u> $\mathcal{T}$ is a species (i.e. property) R of elements of M such that

i) If $a \in R$ and $a \geqslant b$, then $b \in R$.

ii) If $b[^{a}/x]c_1 \ldots c_n \in R$, $n \geqslant 0$ and $a \in M$, then $(\lambda Xb)ac_1 \ldots c_n \in R$.

iii) If $Ka_1 \ldots a_n$ is in M (i.e. $a_1, \ldots, a_n$ are in M), $n \geqslant 0$, then $Ka_1 \ldots a_n \in R$.

An n-ary <u>propositional</u> <u>function</u> <u>over</u> $\mathcal{C}$, $n \geqslant 0$, is a species R of n + 1-tuples $t_1, \ldots, t_n, a$ with $t_1, \ldots, t_n \in I^i$ and $a \in M$ such that for all $t_1, \ldots, t_n \in I^i$, the species

$$Rt_1 \ldots t_n = \{a : \langle t_1, \ldots, t_n, a \rangle \in R\}$$

is a proposition over $\mathcal{C}$ (i.e. satisfies i) - iii)).

We extend $\mathcal{S}$ to the system $\mathcal{T}$ by adding an n-ary relation constant for each n-ary propositional function over $\mathcal{C}$ and a proposition constant for each proposition over $\mathcal{C}$. If P^n is such a constant, then $\overline{P^n}$ is the proposition or propositional function it denotes ($n \geqslant 0$).

I^n denotes the set of all P^n in $\mathcal{T}$.

For each sentence (i.e. closed formula) A of $\mathcal{T}$, we define the species $\overline{A}$ of terms of $\mathcal{C}$ by induction on the number of occurrences of $\supset$ or $\forall$ in A:

$$\overline{P^n t_1 \ldots t_n} = \overline{P^n} t_1 \ldots t_n$$

$$c \in \overline{A \supset B} \equiv \forall a \in \overline{A}\ (ca \quad \overline{B})$$

$$c \in \overline{\forall X^{\alpha} A} \equiv \forall P^{\alpha} \in I^{\alpha} (cK \in \overline{A[^{P}/X^{\alpha}]})$$

<u>Proposition 1</u>. For each sentence A of $\mathcal{T}$, $\overline{A}$ is a proposition over $\mathcal{C}$.

This is clear if A is atomic, i.e. $= P^n t_1 \ldots t_n$. Assume that $\overline{A}$ and $\overline{B}$ are propositions over $\mathcal{C}$ and $c \in \overline{A \supset B}$. $K \in \overline{A}$ and so $cK \in \overline{B}$. Hence, $cK \in M$ and so $c \in M$. Let $c \geqslant d$. Then for all $a \in \overline{A}$, $ca \geqslant da$ and so $da \in \overline{B}$. Hence, $d \in \overline{A \supset B}$. Let $c[^{d}/X]e_1 \ldots e_n \in \overline{A \supset B}$ and $d \in M$. For all $a \in \overline{A}$, $c[^{d}/X]e_1 \ldots e_n a \in \overline{B}$ and so $(\lambda Xc)de_1 \ldots e_n a \in \overline{B}$ by ii). Hence $(\lambda Xc)de_1 \ldots e_n \in \overline{A \supset B}$. Let $Ka_1 \ldots a_n \in M$. Then for all $a \in \overline{A}$, $Ka_1 \ldots a_n a$ is in M and so in $\overline{B}$. Hence, $Ka_1 \ldots a_n \in \overline{A \supset B}$. Thus, $\overline{A \supset B}$ is a proposition over $\mathcal{C}$.

A similar argument proves that $\overline{\forall X^{\alpha} A}$ is a proposition over $\mathcal{C}$ if $A[^{P^{\alpha}}/X^{\alpha}]$ is for each $P^{\alpha} \in I^{\alpha}$.

<u>Qed</u>.

Let T be a closed n-term of $\mathcal{T}$, $n \geq 0$. We define the n-ary propositional function $\overline{T}$ over $\mathcal{C}$ (a proposition if n = 0) by

$$\overline{T}t_1\ldots t_n = \overline{Tt_1\ldots t_n}$$

Thus, there is a relation constant P_T in I^n with $\overline{P}_T = \overline{T}$. By induction on A, it easily follows that

$$(1) \qquad \overline{A[^T/X^n]} = \overline{A[^{P_T}/X^n]}$$

From (1) and the definition of $\overline{\forall X^\alpha A}$, we obtain

$$(2) \qquad c \in \overline{\forall X^\alpha A} \Rightarrow cK \in \overline{A[^T/X^\alpha]}$$

for each closed α-term T of $\mathcal{T}$.

When $a \in \overline{A}$, we say that a <u>realizes</u> A. This is closely related to Kleene's 1945 recursive realizability interpretation, except that, instead of coding functions by their Godel numbers, we use the corresponding term of $\mathcal{C}$.

3.3. It will be convenient to make the purely notational restriction on A-terms of $\mathcal{S}$ that a variable occurs in one of them with at most one superscript. I.e. if X^A and X^B occur in $c \models C$, then A = B; if X^α and X^β occur in c, then α = β, and not both X^A and X^α occur in c.

With each A-term a we associate a term $\overline{a}$ of $\mathcal{C}$ as follows:

$$\overline{X^A} = X$$

$$\overline{\lambda X^A b} = \lambda X\overline{b}$$

$$\overline{\lambda X^\alpha b} = \lambda X\overline{b}$$

$$\overline{ca} = \overline{c}\,\overline{a}$$

$$\overline{cT^\alpha} = \overline{c}K$$

Let $a \models A$ and let $Y_1^{\alpha 1}, \ldots, Y_n^{\alpha n}$ include all the free variables in a of the form Y^{α}. If T_i is a closed α_i-term of $\widetilde{\mathcal{J}}$, set

$$a_0 = a[^T 1/Y_1, \ldots, {}^T n/Y_n]$$

$$A_0 = A[^T 1/Y_1, \ldots, {}^T n/Y_n]$$

Then A_0 is a sentence. Let $Z_1^{B1}, \ldots, Z_m^{Bm}$ be all the free variables in a_0. Then $B_1, \ldots, B_n$ are sentences of $\mathcal{J}$. Let $b_i \in \overline{B}_i$. Then A_0 is called an <u>instance</u> of A and

$$a_1 = \overline{a}_0[^b 1/Z_1, \ldots, {}^b m/Z_m]$$

is called an <u>instance</u> of a for A_0.

<u>Realizability Theorem</u>. If $a \models A$, A_0 is an instance of A and a_1 an instance of a for A_0, then $a_1 \in \overline{A}_0$.

We shall prove this in the next section. As an immediate consequence, we obtain the Well-Foundedness Theorem for $\mathcal{S}$. First, let $a \models A$ and let a (and so A) be closed. Then $\overline{a}$ is an instance of a for A. Hence, $\overline{a}$ is well-founded, by the Realizability Theorem. But, an infinite reduction $a \gg b \gg c \gg \ldots$ would yield an infinite reduction $\overline{a} \gg \overline{b} \gg \overline{c} \gg \ldots$; and so, a is well-founded. Let $a \models A$ and let a be open now. If $X_1^{\alpha 1}, \ldots, X_m^{\alpha m}, Y_1^{B1}, \ldots, Y_n^{Bn}$ are its free variables, then

$$\lambda X_1^{\alpha 1} \ldots \lambda X_m^{\alpha m} \lambda Y_1^{B1} \ldots \lambda Y_n^{Bn}\, a \models$$

$$\forall X_1^{\alpha 1} \ldots X_m^{\alpha m}(B_1 \supset (\ldots \supset (B_n \supset A) \ldots)).$$

But, this term is closed, and hence, well-founded. But, that implies that a is well-founded.

4. Proof of the Realizability Theorem.

4.1. First, we shall prove the Lemma.

Proof. Let

$$(\lambda Xb)ac_1\ldots c_n \gg d_0 \gg d_1 \gg \ldots$$

be a reduction. We must show that it is finite. Note that b is well-founded, since an infinite reduction of b would yield an infinite reduction of $b[^a/x]c_1\ldots c_n$. Likewise, each c_i is well-founded. If each d_i is of the form $(\lambda Xb')a'c_1'\ldots c_n'$ where $b \geqslant b'$, $a \geqslant a'$ and $c_i \geqslant c_i'$, the reduction is finite because b, a, c_1, ..., c_n are well-founded. If some of the d_i are not of this form, then for some j, $d_j = b'[^{a'}/x]$ $c_1'\ldots c_n'$ where $b \geqslant b'$, $a \geqslant a'$, and $c_i \geqslant c_i'$. But then $b[^a/x]\ c_1\ldots c_n \geqslant d_j$, and so the reduction $d_j \geqslant d_{j+1} \geqslant \ldots$ must be finite, since $b[^a/x]c_1\ldots c_n$ is well-founded.

Qed.

4.2. We prove the Realizability Theorem by induction on a.

Case 1. $a = x^A$. Then by definition, $a_1 \in \bar{A}_0$.

Case 2. $a = \lambda x^C b$, where $b \models B$ and so $A = C \supset B$. Let $c \in \bar{C}_0$. Then $b_1[^c/x]$ is an instance of b for B_0; and so by the induction hypothesis, $b_1[^c/x] \in \bar{B}_0$. Hence, by clause ii) in the definition of a proposition over $\mathfrak{C}$, $a_1 c \in \bar{B}_0$. Thus, $a_1 \in \bar{A}_0$.

Case 3. $a = \lambda X^\alpha b$ where $b \models B$ and so $A = \forall X^\alpha B$. Let $P \in I^\alpha$. Then $b_1[^K/X]$ is an instance of b for B_0 (since by the restriction on λX^α, X^α is not free in any C when Y^C is free in b). So, $b_1[^K/X] \in \bar{B}_0$ by the induction hypothesis; and so $a_1 K \in \bar{B}_0$. I.e. $a_1 \in \bar{A}_0$.

Case 4. $a = bc$, where $b \models B \supset A$ and $c \models B$. By the induction hypothesis, $b_1 \in \overline{B_0 \supset C_0}$ and $c_1 \in \bar{B}_0$. So $a_1 = b_1 c_1 \in \bar{A}_0$

Case 5. $a = bT$, where $b \models \forall X^\alpha B$ and $A = B[^T/\ X^\alpha]$. $b_1 \in \overline{(\forall X^\alpha B)}_0$ and so,

$a_1 = b_1 K \in \overline{B[^T/\ X^\alpha]}$ by (2).

Qed.

University of Chicago

References

Church, A. 1932. A set of postulates for the foundations of mathematics. Ann. of Math. (2), 33: 346-366.

Church, A. & Rosser, J.B. 1936. Some properties of conversion. Trans. Amer. Math. Soc. 39: 472-482

Curry, H. & Feys, R. 1958. Combinatory Logic. North Holland, Amsterdam.

Dedekind, R. 1888. Was sind und was sollen die Zahlen? Brunswick.

Gentzen, G. 1934. Untersuchungen uber das logische Schlussen, Mathematische Zeitschrift 39: 176-210, 405-431.

Girard, J.-Y. 1970. Une extension de l'interpretation de Godel a l'analyse et la theorie des types. Proc of the Second Scandanavian Logic Symposium, ed. J.E. Fenstad. North Holland, Amsterdam. 1971. 63-92.

Godel, K. 1932-3. Zur intuitionistischen Arithmetik und Zahlentheorie. Ergebnisse eines math. Koll. Heft 4. 34-38.

Martin-Lof, P. 1970. Hauptsatz for the theory of species. Proc of the Second Scandanavian Logic Symposium, ed. J.E. Fenstad. North Holland, Amsterdam. 1971. 217-233.

Prawitz, D. 1965. Natural Deduction. Almqvist and Wiksell. Stockholm.

__________. 1968. Some results for intuitionistic logic with second order quantification rules. Intuitionism and Proof Theory. eds. A. Kino, J. Myhill, R.E. Vesley. North Holland, Amsterdam. 1970.

__________. 1970. Ideas and results of proof theory. Proc of the Second Scandinavian Logic Symposium, ed. J.E. Fenstad. North Holland, Amsterdam. 1971. 235-307.

Vol. 277: Séminaire Banach. Edité par C. Houzel. VII, 229 pages. 1972. DM 22,–

Vol. 278: H. Jacquet, Automorphic Forms on GL(2) Part II. XIII, 142 pages. 1972. DM 18,–

Vol. 279: R. Bott, S. Gitler and I. M. James, Lectures on Algebraic and Differential Topology. V, 174 pages. 1972. DM 20,–

Vol. 280: Conference on the Theory of Ordinary and Partial Differential Equations. Edited by W. N. Everitt and B. D. Sleeman. XV, 367 pages. 1972. DM 29,–

Vol. 281: Coherence in Categories. Edited by S. Mac Lane. VII, 235 pages. 1972. DM 22,–

Vol. 282: W. Klingenberg und P. Flaschel, Riemannsche Hilbertmannigfaltigkeiten. Periodische Geodätische. VII, 211 Seiten. 1972. DM 22,–

Vol. 283: L. Illusie, Complexe Cotangent et Déformations II. VII, 304 pages. 1972. DM 27,–

Vol. 284: P. A. Meyer, Martingales and Stochastic Integrals I. VI, 89 pages. 1972. DM 18,–

Vol. 285: P. de la Harpe, Classical Banach-Lie Algebras and Banach-Lie Groups of Operators in Hilbert Space. III, 160 pages. 1972. DM 18,–

Vol. 286: S. Murakami, On Automorphisms of Siegel Domains. V, 95 pages. 1972. DM 18,–

Vol. 287: Hyperfunctions and Pseudo-Differential Equations. Edited by H. Komatsu. VII, 529 pages. 1973. DM 40,–

Vol. 288: Groupes de Monodromie en Géométrie Algébrique. (SGA 7 I). Dirigé par A. Grothendieck. IX, 523 pages. 1972. DM 55,–

Vol. 289: B. Fuglede, Finely Harmonic Functions. III, 188. 1972. DM 20,–

Vol. 290: D. B. Zagier, Equivariant Pontrjagin Classes and Applications to Orbit Spaces. IX, 130 pages. 1972. DM 18,–

Vol. 291: P. Orlik, Seifert Manifolds. VIII, 155 pages. 1972. DM 18,–

Vol. 292: W. D. Wallis, A. P. Street and J. S. Wallis, Combinatorics: Room Squares, Sum-Free Sets, Hadamard Matrices. V, 508 pages. 1972. DM 55,–

Vol. 293: R. A. DeVore, The Approximation of Continuous Functions by Positive Linear Operators. VIII, 289 pages. 1972. DM 27,–

Vol. 294: Stability of Stochastic Dynamical Systems. Edited by R. F. Curtain. IX, 332 pages. 1972. DM 29,–

Vol. 295: C. Dellacherie, Ensembles Analytiques Capacités. Mesures de Hausdorff. XII, 123 pages. 1972. DM 18,–

Vol. 296: Probability and Information Theory II. Edited by M. Behara, K. Krickeberg and J. Wolfowitz. V, 223 pages. 1973. DM 22,–

Vol. 297: J. Garnett, Analytic Capacity and Measure. IV, 138 pages. 1972. DM 18,–

Vol. 298: Proceedings of the Second Conference on Compact Transformation Groups. Part 1. XIII, 453 pages. 1972. DM 35,–

Vol. 299: Proceedings of the Second Conference on Compact Transformation Groups. Part 2. XIV, 327 pages. 1972. DM 29,–

Vol. 300: P. Eymard, Moyennes Invariantes et Représentations Unitaires. II, 113 pages. 1972. DM 18,–

Vol. 301: F. Pittnauer, Vorlesungen über asymptotische Reihen. VI, 186 Seiten. 1972. DM 18,–

Vol. 302: M. Demazure, Lectures on p-Divisible Groups. V, 98 pages. 1972. DM 18,–

Vol. 303: Graph Theory and Applications. Edited by Y. Alavi, D. R. Lick and A. T. White. IX, 329 pages. 1972. DM 26,–

Vol. 304: A. K. Bousfield and D. M. Kan, Homotopy Limits, Completions and Localizations. V, 348 pages. 1972. DM 29,–

Vol. 305: Théorie des Topos et Cohomologie Etale des Schémas. Tome 3. (SGA 4). Dirigé par M. Artin, A. Grothendieck et J. L. Verdier. VI, 640 pages. 1973. DM 55,–

Vol. 306: H. Luckhardt, Extensional Gödel Functional Interpretation. VI, 161 pages. 1973. DM 20,–

Vol. 307: J. L. Bretagnolle, S. D. Chatterji et P.-A. Meyer, Ecole d'été de Probabilités: Processus Stochastiques. VI, 198 pages. 1973. DM 22,–

Vol. 308: D. Knutson, λ-Rings and the Representation Theory of the Symmetric Group. IV, 203 pages. 1973. DM 22,–

Vol. 309: D. H. Sattinger, Topics in Stability and Bifurcation Theory. VI, 190 pages. 1973. DM 20,–

Vol. 310: B. Iversen, Generic Local Structure of the Morphisms in Commutative Algebra. IV, 108 pages. 1973. DM 18,–

Vol. 311: Conference on Commutative Algebra. Edited by J. W. Brewer and E. A. Rutter. VII, 251 pages. 1973. DM 24,–

Vol. 312: Symposium on Ordinary Differential Equations. Edited by W. A. Harris, Jr. and Y. Sibuya. VIII, 204 pages. 1973. DM 22,–

Vol. 313: K. Jörgens and J. Weidmann, Spectral Properties of Hamiltonian Operators. III, 140 pages. 1973. DM 18,–

Vol. 314: M. Deuring, Lectures on the Theory of Algebraic Functions of One Variable. VI, 151 pages. 1973. DM 18,–

Vol. 315: K. Bichteler, Integration Theory (with Special Attention to Vector Measures). VI, 357 pages. 1973. DM 29,–

Vol. 316: Symposium on Non-Well-Posed Problems and Logarithmic Convexity. Edited by R. J. Knops. V, 176 pages. 1973. DM 20,–

Vol. 317: Séminaire Bourbaki – vol. 1971/72. Exposés 400–417. IV, 361 pages. 1973. DM 29,–

Vol. 318: Recent Advances in Topological Dynamics. Edited by A. Beck. VIII, 285 pages. 1973. DM 27,–

Vol. 319: Conference on Group Theory. Edited by R. W. Gatterdam and K. W. Weston. V, 188 pages. 1973. DM 20,–

Vol. 320: Modular Functions of One Variable I. Edited by W. Kuyk. V, 195 pages. 1973. DM 20,–

Vol. 321: Séminaire de Probabilités VII. Edité par P. A. Meyer. VI, 322 pages. 1973. DM 29,–

Vol. 322: Nonlinear Problems in the Physical Sciences and Biology. Edited by I. Stakgold, D. D. Joseph and D. H. Sattinger. VIII, 357 pages. 1973. DM 29,–

Vol. 323: J. L. Lions, Perturbations Singulières dans les Problèmes aux Limites et en Contrôle Optimal. XII, 645 pages. 1973. DM 46,–

Vol. 324: K. Kreith, Oscillation Theory. VI, 109 pages. 1973. DM 18,–

Vol. 325: C.-C. Chou, La Transformation de Fourier Complexe et L'Equation de Convolution. IX, 137 pages. 1973. DM 18,–

Vol. 326: A. Robert, Elliptic Curves. VIII, 264 pages. 1973. DM 24,–

Vol. 327: E. Matlis, One-Dimensional Cohen-Macaulay Rings. XII, 157 pages. 1973. DM 20,–

Vol. 328: J. R. Büchi and D. Siefkes, The Monadic Second Order Theory of All Countable Ordinals. VI, 217 pages. 1973. DM 22,–

Vol. 329: W. Trebels, Multipliers for (C, α)-Bounded Fourier Expansions in Banach Spaces and Approximation Theory. VII, 103 pages. 1973. DM 18,–

Vol. 330: Proceedings of the Second Japan-USSR Symposium on Probability Theory. Edited by G. Maruyama and Yu. V. Prokhorov. VI, 550 pages. 1973. DM 40,–

Vol. 331: Summer School on Topological Vector Spaces. Edited by L. Waelbroeck. VI, 226 pages. 1973. DM 22,–

Vol. 332: Séminaire Pierre Lelong (Analyse) Année 1971-1972. V, 131 pages. 1973. DM 18,–

Vol. 333: Numerische, insbesondere approximationstheoretische Behandlung von Funktionalgleichungen. Herausgegeben von R. Ansorge und W. Törnig. VI, 296 Seiten. 1973. DM 27,–

Vol. 334: F. Schweiger, The Metrical Theory of Jacobi-Perron Algorithm. V, 111 pages. 1973. DM 18,–

Vol. 335: H. Huck, R. Roitzsch, U. Simon, W. Vortisch, R. Walden, B. Wegner und W. Wendland, Beweismethoden der Differentialgeometrie im Großen. IX, 159 Seiten. 1973. DM 20,–

Vol. 336: L'Analyse Harmonique dans le Domaine Complexe. Edité par E. J. Akutowicz. VIII, 169 pages. 1973. DM 20,–

Vol. 337: Cambridge Summer School in Mathematical Logic. Edited by A. R. D. Mathias and H. Rogers. IX, 660 pages. 1973. DM 46,–

Vol: 338: J. Lindenstrauss and L. Tzafriri, Classical Banach Spaces. IX, 243 pages. 1973. DM 24,–

Vol. 339: G. Kempf, F. Knudsen, D. Mumford and B. Saint-Donat, Toroidal Embeddings I. VIII, 209 pages. 1973. DM 22,–

Vol. 340: Groupes de Monodromie en Géométrie Algébrique. (SGA 7 II). Par P. Deligne et N. Katz. X, 438 pages. 1973. DM 44,–

Vol. 341: Algebraic K-Theory I, Higher K-Theories. Edited by H. Bass. XV, 335 pages. 1973. DM 29,–

Vol. 342: Algebraic K-Theory II, "Classical" Algebraic K-Theory, and Connections with Arithmetic. Edited by H. Bass. XV, 527 pages. 1973. DM 40,–

Vol. 343: Algebraic K-Theory III, Hermitian K-Theory and Geometric Applications. Edited by H. Bass. XV, 572 pages. 1973. DM 40,–

Vol. 344: A. S. Troelstra (Editor), Metamathematical Investigation of Intuitionistic Arithmetic and Analysis. XVII, 485 pages. 1973. DM 38,–

Vol. 345: Proceedings of a Conference on Operator Theory. Edited by P. A. Fillmore. VI, 228 pages. 1973. DM 22,–

Vol. 346: Fučík et al., Spectral Analysis of Nonlinear Operators. II, 287 pages. 1973. DM 26,–

Vol. 347: J. M. Boardman and R. M. Vogt, Homotopy Invariant Algebraic Structures on Topological Spaces. X, 257 pages. 1973. DM 24,–

Vol. 348: A. M. Mathai and R. K. Saxena, Generalized Hypergeometric Functions with Applications in Statistics and Physical Sciences. VII, 314 pages. 1973. DM 26,–

Vol. 349: Modular Functions of One Variable II. Edited by W. Kuyk and P. Deligne. V, 598 pages. 1973. DM 38,–

Vol. 350: Modular Functions of One Variable III. Edited by W. Kuyk and J.-P. Serre. V, 350 pages. 1973. DM 26,–

Vol. 351: H. Tachikawa, Quasi-Frobenius Rings and Generalizations. XI, 172 pages. 1973. DM 20,–

Vol. 352: J. D. Fay, Theta Functions on Riemann Surfaces. V, 137 pages. 1973. DM 18,–

Vol. 353: Proceedings of the Conference on Orders, Group Rings and Related Topics. Organized by J. S. Hsia, M. L. Madan and T. G. Ralley. X, 224 pages. 1973. DM 22,–

Vol. 354: K. J. Devlin, Aspects of Constructibility. XII, 240 pages. 1973. DM 24,–

Vol. 355: M. Sion, A Theory of Semigroup Valued Measures. V, 140 pages. 1973. DM 18,–

Vol. 356: W. L. J. van der Kallen, Infinitesimally Central-Extensions of Chevalley Groups. VII, 147 pages. 1973. DM 18,–

Vol. 357: W. Borho, P. Gabriel und R. Rentschler, Primideale in Einhüllenden auflösbarer Lie-Algebren. V, 182 Seiten. 1973. DM 20,–

Vol. 358: F. L. Williams, Tensor Products of Principal Series Representations. VI, 132 pages. 1973. DM 18,–

Vol. 359: U. Stammbach, Homology in Group Theory. VIII, 183 pages. 1973. DM 20,–

Vol. 360: W. J. Padgett and R. L. Taylor, Laws of Large Numbers for Normed Linear Spaces and Certain Fréchet Spaces. VI, 111 pages. 1973. DM 18,–

Vol. 361: J. W. Schutz, Foundations of Special Relativity: Kinematic Axioms for Minkowski Space Time. XX, 314 pages. 1973. DM 26,–

Vol. 362: Proceedings of the Conference on Numerical Solution of Ordinary Differential Equations. Edited by D. Bettis. VIII, 490 pages. 1974. DM 34,–

Vol. 363: Conference on the Numerical Solution of Differential Equations. Edited by G. A. Watson. IX, 221 pages. 1974. DM 20,–

Vol. 364: Proceedings on Infinite Dimensional Holomorphy. Edited by T. L. Hayden and T. J. Suffridge. VII, 212 pages. 1974. DM 20,–

Vol. 365: R. P. Gilbert, Constructive Methods for Elliptic Equations. VII, 397 pages. 1974. DM 26,–

Vol. 366: R. Steinberg, Conjugacy Classes in Algebraic Groups (Notes by V. V. Deodhar). VI, 159 pages. 1974. DM 18,–

Vol. 367: K. Langmann und W. Lütkebohmert, Cousinverteilungen und Fortsetzungssätze. VI, 151 Seiten. 1974. DM 16,–

Vol. 368: R. J. Milgram, Unstable Homotopy from the Stable Point of View. V, 109 pages. 1974. DM 16,–

Vol. 369: Victoria Symposium on Nonstandard Analysis. Edited by A. Hurd and P. Loeb. XVIII, 339 pages. 1974. DM 26,–

Vol. 370: B. Mazur and W. Messing, Universal Extensions and One Dimensional Crystalline Cohomology. VII, 134 pages. 1974. DM 16,–

Vol. 371: V. Poenaru, Analyse Différentielle. V, 228 pages. 1974. DM 20,–

Vol. 372: Proceedings of the Second International Conference on the Theory of Groups 1973. Edited by M. F. Newman. VII, 740 pages. 1974. DM 48,–

Vol. 373: A. E. R. Woodcock and T. Poston, A Geometrical Study of the Elementary Catastrophes. V, 257 pages. 1974. DM 22,–

Vol. 374: S. Yamamuro, Differential Calculus in Topological Linear Spaces. IV, 179 pages. 1974. DM 18,–

Vol. 375: Topology Conference 1973. Edited by R. F. Dickman Jr. and P. Fletcher. X, 283 pages. 1974. DM 24,–

Vol. 376: D. B. Osteyee and I. J. Good, Information, Weight of Evidence, the Singularity between Probability Measures and Signal Detection. XI, 156 pages. 1974. DM 16,–

Vol. 377: A. M. Fink, Almost Periodic Differential Equations. VIII, 336 pages. 1974. DM 26,–

Vol. 378: TOPO 72 – General Topology and its Applications. Proceedings 1972. Edited by R. Alò, R. W. Heath and J. Nagata. XIV, 651 pages. 1974. DM 50,–

Vol. 379: A. Badrikian et S. Chevet, Mesures Cylindriques, Espaces de Wiener et Fonctions Aléatoires Gaussiennes. X, 383 pages. 1974. DM 32,–

Vol. 380: M. Petrich, Rings and Semigroups. VIII, 182 pages. 1974. DM 18,–

Vol. 381: Séminaire de Probabilités VIII. Edité par P. A. Meyer. IX, 354 pages. 1974. DM 32,–

Vol. 382: J. H. van Lint, Combinatorial Theory Seminar Eindhoven University of Technology. VI, 131 pages. 1974. DM 18,–

Vol. 383: Séminaire Bourbaki – vol. 1972/73. Exposés 418-435 IV, 334 pages. 1974. DM 30,–

Vol. 384: Functional Analysis and Applications, Proceedings 1972. Edited by L. Nachbin. V, 270 pages. 1974. DM 22,–

Vol. 385: J. Douglas Jr. and T. Dupont, Collocation Methods for Parabolic Equations in a Single Space Variable (Based on C^1-Piecewise-Polynomial Spaces). V, 147 pages. 1974. DM 16,–

Vol. 386: J. Tits, Buildings of Spherical Type and Finite BN-Pairs. IX, 299 pages. 1974. DM 24,–

Vol. 387: C. P. Bruter, Eléments de la Théorie des Matroïdes. V, 138 pages. 1974. DM 18,–

Vol. 388: R. L. Lipsman, Group Representations. X, 166 pages. 1974. DM 20,–

Vol. 389: M.-A. Knus et M. Ojanguren, Théorie de la Descente et Algèbres d' Azumaya. IV, 163 pages. 1974. DM 20,–

Vol. 390: P. A. Meyer, P. Priouret et F. Spitzer, Ecole d'Eté de Probabilités de Saint-Flour III – 1973. Edité par A. Badrikian et P.-L. Hennequin. VIII, 189 pages. 1974. DM 20,–

Vol. 391: J. Gray, Formal Category Theory: Adjointness for 2-Categories. XII, 282 pages. 1974. DM 24,–

Vol. 392: Géométrie Différentielle, Colloque, Santiago de Compostela, Espagne 1972. Edité par E. Vidal. VI, 225 pages. 1974. DM 20,–

Vol. 393: G. Wassermann, Stability of Unfoldings. IX, 164 pages. 1974. DM 20,–

Vol. 394: W. M. Patterson 3rd, Iterative Methods for the Solution of a Linear Operator Equation in Hilbert Space – A Survey. III, 183 pages. 1974. DM 20,–

Vol. 395: Numerische Behandlung nichtlinearer Integrodifferential- und Differentialgleichungen. Tagung 1973. Herausgegeben von R. Ansorge und W. Törnig. VII, 313 Seiten. 1974. DM 28,–

Vol. 396: K. H. Hofmann, M. Mislove and A. Stralka, The Pontryagin Duality of Compact O-Dimensional Semilattices and its Applications. XVI, 122 pages. 1974. DM 18,–

Vol. 397: T. Yamada, The Schur Subgroup of the Brauer Group. V, 159 pages. 1974. DM 18,–

Vol. 398: Théories de l'Information, Actes des Rencontres de Marseille-Luminy, 1973. Edité par J. Kampé de Fériet et C. Picard. XII, 201 pages. 1974. DM 23,–